# Building and Designing Decks

# Building and Designing Decks

SCOTT SCHUTTNER

The Taunton Press

Cover photos: Chuck Miller
Text photos: Jeff Beneke, except where noted

Printed in the United States of America
10 9 8 7 6 5 4


For Pros / By Pros®: Building and Designing Decks
was originally published in 1993 by
The Taunton Press, Inc.

For Pros / By Pros® is a trademark of The Taunton Press, Inc.,
registered in the U.S. Patent and Trademark Office.

**The Taunton Press**
Inspiration for hands-on living®

The Taunton Press, Inc., 63 South Main Street,
PO Box 5506, Newtown, CT 06470-5506
e-mail: tp@taunton.com




Library of Congress Cataloging-in-Publication Data

Schuttner, Scott, 1947
For Pros / By Pros®: Building and Designing Decks /
Scott Schuttner.
p.   cm.
Includes index.
ISBN 1-56158-320-0
1. Decks (Architecture, Domestic) — Design and Construction.  I. Title
TH4970.S39    1998              92-30687
690'.89 — dc20                  CIP

This book is dedicated to my mother, father and sister in Oklahoma, so very far away.

# Contents

# Introduction

A deck is somewhat like a bridge. It connects us to two separate spaces while safely elevating us above ground level. People not only enjoy relaxing and playing on decks, they also just plain like being on decks.

Why add a deck? The most common reason is to create a place to play outdoors in nice weather. If the site is too rocky, steep or wet, a deck creates a much appreciated level surface. Maybe you're adding on to improve the looks or salability of your home. Once you've honestly evaluated your objectives, it will be easier to make some good choices on the kind of deck you need.

The concept of a transition area between inside and outside has a long history in architecture. The term "porch" comes from the Greek word *portico,* which means a roof with columns. Porches provide protection from the sun and rain, but a century ago they also served as a connection to the social life of the community. But as the value of individual privacy grew, and as the automobile made street watching less appealing, we moved our transition area from the front of the house to the back. Here, the porch often evolved into the patio, which also served an important social function, but in a more limited sense.

Concrete patios do have some redeeming qualities. They're relatively inexpensive, and if built properly they're maintenance free and very durable—maybe a little too durable. But, for many people, they're just plain ugly. We see so much pavement all around us that we prefer not to have to relax on it as well. A wood deck has a much more relaxing and "natural" feel to it.

Why decks and not porches? Decks first flourished in areas with consistently nice climates, such as California. They were considered to be more modern and this often appealed to designers who were trying to break from tradition. They were easier to add onto existing homes. Decks got us out into the sunshine, which, at the time, seemed like the best place to be. Above all, decks are simpler and easier to build than porches, and therefore less expensive. They don't need a roof and all its complications, and so they require less material and labor. This dollar savings is often negated, however, as people build decks that are much larger and more elaborate than any porch would ever be. But decks can be simple, and that makes them ideal projects for the home owner and do-it-yourselfer. Porches are being revived as we search for tradition in American architecture, but decks are here to stay.

**Acknowledgments** I'd like to thank Jim Hall for providing the initial impetus for this book and video project. Thanks also to my fellow carpenters and our clients, who allowed me time off from my other responsibilities while I was writing. Credit goes to my editor, Jeff Beneke, for his help in refining my ideas and suggesting new ones. Thanks to Tom Law, for his careful reading of the manuscript and useful suggestions. Finally, thanks to Glynn, Silvan and Linnaea for waiting patiently.

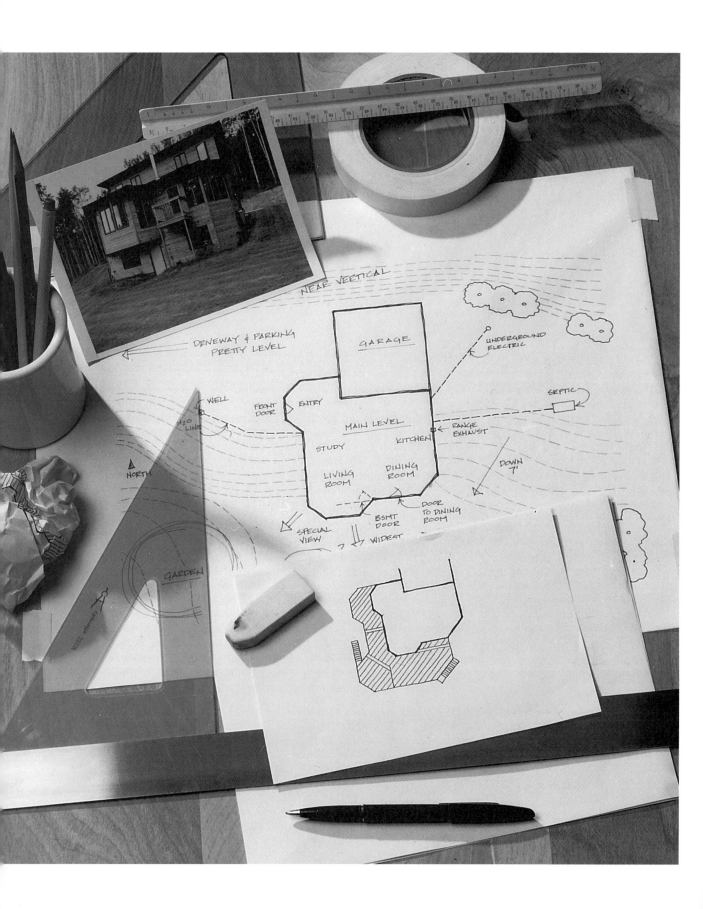

NEAR VERTICAL

DRIVEWAY & PARKING
PRETTY LEVEL

GARAGE

UNDERGROUND
ELECTRIC

WELL

FRONT
DOOR

ENTRY

H₂O
LINE

MAIN LEVEL

SEPTIC

RANGE
EXHAUST

STUDY

KITCHEN

NORTH

LIVING
ROOM

DINING
ROOM

DOWN
7'

SPECIAL
VIEW

BSMT
DOOR

DOOR
TO DINING
ROOM

WIDEST

GARDEN

# Designing a Deck

## Chapter 1

A good deck requires quality materials and construction, but it begins with a good design. Design plans are important for several reasons. They enhance communication among everyone involved in the project. A set of plans is the common reference for the client, the designer and the builder. Even if you're wearing all three of these hats yourself you'll want a reliable set of plans, and if someone else is doing the building, the plans will be your insurance that what gets built is what you want. Plans help minimize guessing and subjective assumptions. Also, you're probably going to need a building permit to build a deck, and to get that permit, the local building inspectors will want to look at the plans. They'll want to see that the design meets their requirements, and they're going to look more favorably on clear and concise plans.

The design stage forces everyone to take a variety of conflicting or vague ideas about the project and turn them into viable intentions. You don't want to build a deck to see if it meets your expectations—that decision should be made using a pencil and paper and some mental energy.

Designing first on paper will identify problems—finding the best way to connect two parts or working out the amount of room needed for a stairway will go a long way toward making sure that all of the parts will fit. You won't eliminate all the problems, of course, but the idea is to go as far as you can beforehand to minimize mistakes later on.

Design work is a prerequisite for accurate cost estimating. It can be risky to have a fixed budget and begin building a project that still has a lot of undetermined details. Builders often do this (myself included),

but unless they're very clear about who is responsible for what, it will inevitably lead to misunderstandings and bad feelings. Even if you're doing all the labor yourself, your design will help you determine if you can afford the project.

Who should do the design work? I suppose there are some people who don't want to be involved with the design of their home or any improvements to it, but not my clients, and certainly not you if you're reading this book. For a design to work at its best, it must include input from the final users—and the more the better. My favorite clients are the ones who are armed to the teeth with pictures and thoughts of how they think things should work. Granted, these aren't necessarily the easiest jobs to design, but when the dust settles the designs are much more likely to express and fulfill the client's true wishes rather than my perception of what they wanted.

Builders frequently offer design services, and these can be a good value. But there's no guarantee that a successful builder is going to be a reliable designer. Look for recommendations from friends and get references from the builder. If possible, go look at some of the builder's recent jobs. If you're a builder, offer references to prospective clients and provide lots of photos of past work.

Architects are an obvious source for deck designs, but an architectural degree is no guarantee of reliability either, so choose an architect with the same care you use in selecting a builder. Some architects are better at designing shopping malls than decks. An architect may appear to be the most expensive option for a designer, but for a complex deck or a deck that needs to blend with an unusual house, an architect may be the best choice.

If it's going to be your deck, why not design it yourself? For a relatively straightforward deck, if you feel competent enough to build it yourself, you may be the best designer possible. Deck building is a great way to learn some basic design and construction skills, and no one knows better than you what you want and expect from your deck. The bookstores and building-supply stores are full of deck plans, but if you build a deck from a plan prepared by some committee that never looked at your house or asked what you want, you're bound to be less than pleased. Most decks that look as if they were stuck onto a house probably were.

Not everyone has the temperament to do their own design work. It takes a certain facility with the mental manipulation of three-dimensional ideas. This comes naturally to some and may require a lot of practice for others. If you feel overwhelmed by the prospect, or really find your joy only in the actual building, then maybe you'd be happiest letting someone else do the designing. If that's your choice, then this chapter should help you guide your designer toward the deck you want.

One thing is certain. Whether things are worked out in detail way in advance (the ideal) or decided at the last minute (regrettable), someone will have designed the project. Some people think they'll save time and money by solving design problems as they arise, but this is false economy. Every step requires a decision, and the best decisions are the result of foresight and careful planning.

## The Design Process

Design is sometimes said to be the weaving of an aesthetically pleasing composition around practical requirements. Form follows function. Many projects have been built according to this adage, but with mixed or even disastrous results. Why? One reason is because the fullest scope of the word "function" wasn't explored. The most satisfying designs take the meaning of "functional" a step further by placing an emphasis on how the design performs as a system, not just how each part does its job. It's important to keep in mind that the deck does not exist in isolation. A new deck can alter the dynamics among house, lawn and neighbors.

Although performance as a system requires a continuity between parts, this by no means dictates a boring design. As you explore the interrelated parts of a fully functioning design, you'll discover problems re-quiring creative and unique solutions. These problems may suggest relocating or separating large parts of the deck, adding entrances and exits to the deck and the house or changing the relative heights of different areas of the deck.

Design isn't all grand theory, of course. God is in the details, they say, and the small details of design (color, repeating form and proportion, for example), are essential to success when placed in the context of a systematic approach.

The design process is just that—a process. The final design starts from a core idea that gradually evolves through a series of accepted and rejected accessories. Take the core idea and lead it in several directions, then look back and see if any of your efforts make sense.

The best design will consider all the possibilities and then rank the priorities. Some elements will be more important than others. For example, you may want a deck that allows maximum sunlight, which would dictate that it go on the south side of the house for most people. But there's no door on the south side of the house, and maybe a favorite flower garden is already located there. You could add a door, but perhaps that would be beyond your budget. A walkway from an existing door might be a more attractive option. The walkway could pass around or through the flower garden, thus becoming an asset rather than a second-best solution.

That's one direction the design could take. Another would be to split the deck into smaller units that frame the garden, for example, yet allow you to keep most of it. Or the walkway could be enlarged to serve as one of the small deck units. Each solution creates more possibilities as well as problems, some of which can be overcome and some not.

These compromises can be frustrating, and you may want to set the design aside for a while to refresh your perspective. If you're having problems, it's important to define what's making you unhappy with a design. Often just changing one or two details will relieve you of the preoccupation that the whole design was faulty. A good design will call attention to its strengths and naturalness, while a bad design will keep calling attention to its weaknesses.

# Designing the Morris Family's Deck

This is the south view of a house I built for the Morris family in 1989. They wanted a large deck on the house, but at the time could only afford the small walkout off the dining area (shown here). The evolution of the design for the full deck (finally built in 1991) is shown in the drawings below and on the following pages.

*The deck design evolves over time out of some core ideas and a lot of sketches.*

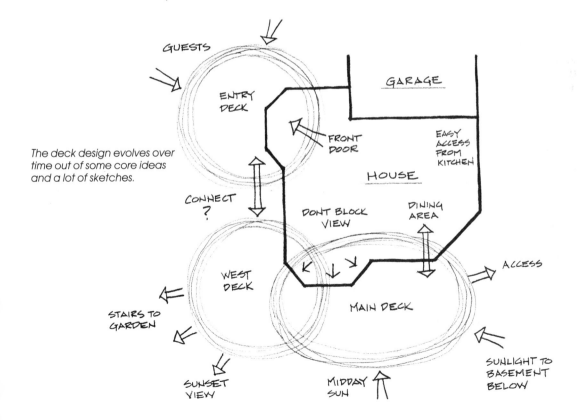

*1. This is the first try at translating the rough ideas of the clients into a real outline of a deck. Although it addresses most of the owners' requisites, it does little else. Unfortunately, too many designs stop at this point.*

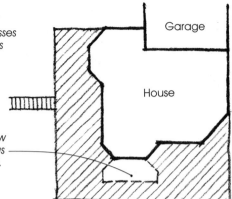

*An opening in the deck here to allow sunlight into the lower basement was considered, but was finally rejected.*

*2. Cutting the corners of the deck at 45° angles is a major stride toward matching the style of the house.*

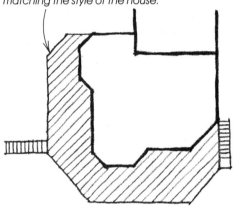

*3. Reducing this area emphasizes the distinction between the main entry and the west deck. It now serves as a walkway rather than as a full section of the deck. One big problem still remains—the main area of the deck interferes with the southerly view from inside the house.*

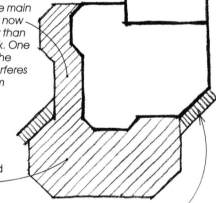

*This area was increased to provide room for a future hot tub.*

*Cutting back this section allows a view out of the southeast windows.*

*4. The next step separates the deck into three distinct levels (A, B and C). Level A is at door height for the entry, Level B is 15 in. below Level A and Level C is 22½ in. below Level B. This accomplishes several objectives: it allows a better view of the yard from inside the house, it complements the downward slope of the land and it prevents the deck from towering way above the ground, actually drawing the lawn and deck a bit closer together.*

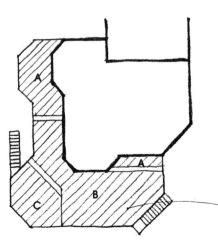

*Reducing the size of this area saves some money.*

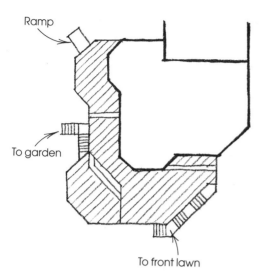

Ramp

To garden

To front lawn

*5. The final step in the design process directs the stairways into the front lawn and garden areas, and a ramp was added near the front door. Rather than construct built-in seating, which might block the view, the level changes were broadened to make them useful for seating.*

**As you can see in this wide-angle photo, the finished deck is more than a simple 'add-on' to the house. It is in fact a significant architectural addition to the house. The deck doesn't simply allow the occupants to lounge around outside the house, it invites them to reorient the way they live in and interact with their immediate surroundings. The results of a thoughtful design can be many years of enjoyment. (Photo by Jim Hall)**

## How to Design

The design process begins by listing the reasons for wanting a deck. Some may be pretty obvious and others may not become apparent until you've given the matter some thought over a period of time or lived in the house for a while. If the deck is being built onto a new house, this prolonged process probably won't be possible. But if you've got the time, let the design grow out of your thoughts and observations. Take notes and make lots of sketches.

The list of things you expect from the deck might well begin with wanting a comfortable outdoor spot to play and relax. The deck may provide a safe, easily enclosed space for kids to play with their blocks or ride a tricycle. Maybe you want to cover up a crumbling patio or hide an ugly fuel tank and create a place to store firewood. The list should address whether you want the deck to be a private retreat where you can drink your morning coffee or a public gathering spot for friends and neighbors—or both. Is the deck needed to make a small house feel bigger or is it more necessary to provide a setting for weekend social life? Are your parties small and intimate or do you entertain large numbers of folks at one time? Is your main goal to improve the appearance and value of your house? Do you want to be able to enjoy the deck when it rains? Do you want to have any lights on it?

This information can quickly start filling up pages of paper, but it's essential to the designer. Once the list feels reasonably complete, start establishing priorities: What's most important? What are you willing to compromise on?

The next step is to convert your list of requirements into some sketches. You might find this easier to do if you take some photographs of the house from several different angles and then enlarge the prints on a copying machine. As you work on sketches for the plan drawings, you can draw the ideas onto the photo images to see if they "fit." These first sketches are the beginning of plan drawings, so work on a large sheet of paper and use a pencil because you'll do a lot of erasing. I like to start with some broad strokes, including the major influences on the deck (house, neighbors, trees, swimming pool or hot tub, street, etc.). Then add circles indicating specific purposes of the deck. Add arrows indicating traffic flow, entrances and exits. Include reminders of such problems as utilities. Soon you'll notice the outline of the deck starting to take shape as you add and subtract ideas. Large main areas will take on clearer definition, while smaller items, such as stairs, begin appearing on the periphery. The initial sketches now need to be given hard edges. You

## A Contrast in Designs

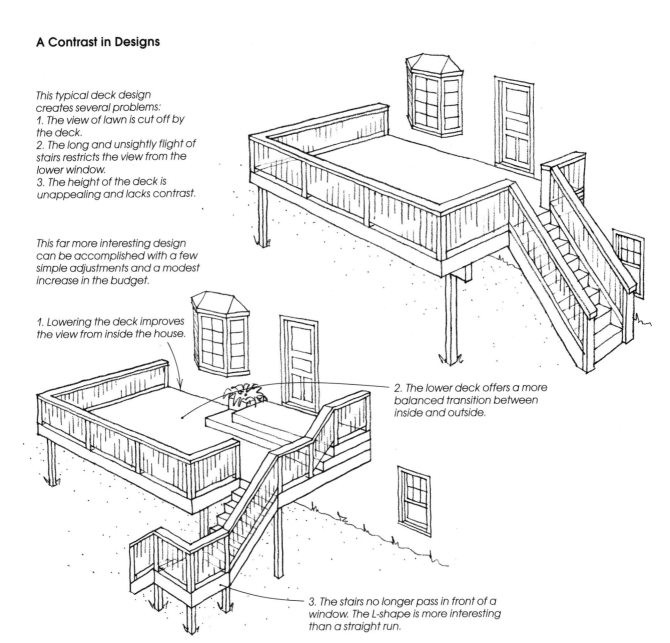

*This typical deck design creates several problems:*
*1. The view of lawn is cut off by the deck.*
*2. The long and unsightly flight of stairs restricts the view from the lower window.*
*3. The height of the deck is unappealing and lacks contrast.*

*This far more interesting design can be accomplished with a few simple adjustments and a modest increase in the budget.*

*1. Lowering the deck improves the view from inside the house.*

*2. The lower deck offers a more balanced transition between inside and outside.*

*3. The stairs no longer pass in front of a window. The L-shape is more interesting than a straight run.*

need to start drawing a real outline with real dimensions. Only then should you start sketching in special features such as level changes, built-in seating and stairs.

Once I've come up with a design that I like (and that my clients like), I start to figure out how to build it. Unless the design is really wild, I can usually make it work. Sometimes, however, the structural solutions to my architectural fantasies are too costly or time consuming, and I have to revise the design—or the budget—if the client is willing. If the design starts appearing to be well beyond what the budget allows, there are several options. You can eliminate some costly features or build a smaller deck. Or you can design the deck so that it can be built in stages. Build the basic deck now and add those special features later on when the client can afford them. If you take this approach, you should start with a design of the deck you want ultimately, which gives you a realistic goal, and then work backwards.

## Design Considerations

Determining what an individual may want from a deck is a subjective experience, by definition. But there are a host of more routine, objective factors that need to be considered in every deck design.

**Climate and weather** Inside the house you can control the temperature and light with a flip of a switch. Not so with decks, and that's the whole idea. Decks serve as a transition area between the environment we maintain inside our homes and the external world. As such, you must consider how the external elements will affect your use and enjoyment of the deck. The position of the deck in relation to the daily path of the sun is one of the most critical considerations. If the deck is too hot or cool to use comfortably, then it won't get used even if it has a spectacular view or is beautifully landscaped. Here in southern Alaska, my clients make the most use of

their decks from late spring to early fall. In northern climates it's common to build the deck on the south side of the home to maximize exposure to the sun. Maintaining a clear western exposure is helpful to capture the evening sun. Shade from the house or surrounding trees might well be unwelcome.

In warm climates, the deck may see year-round use, and the afternoon and early evening sun may be intense enough that your primary goal is to shade the deck. This may mean that the deck would best be placed on a more easterly side of the house, which would provide shade in late afternoon and early evening. In particularly hot climates, a deck on the northern side of the house might be best.

A deck that wraps around a corner of the house can simultaneously provide sun and shade, or you can depend on fences or shrubbery for shade. Movable screens can shade small areas, while an overhead trellis or awning can work wonders. A well-designed trellis should provide shade from the sun when it's at its

## Tracking the Sun

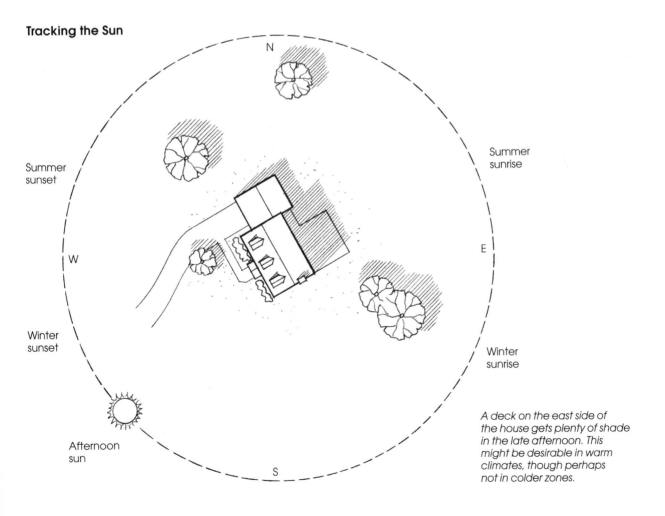

*A deck on the east side of the house gets plenty of shade in the late afternoon. This might be desirable in warm climates, though perhaps not in colder zones.*

## The Trellis Effect

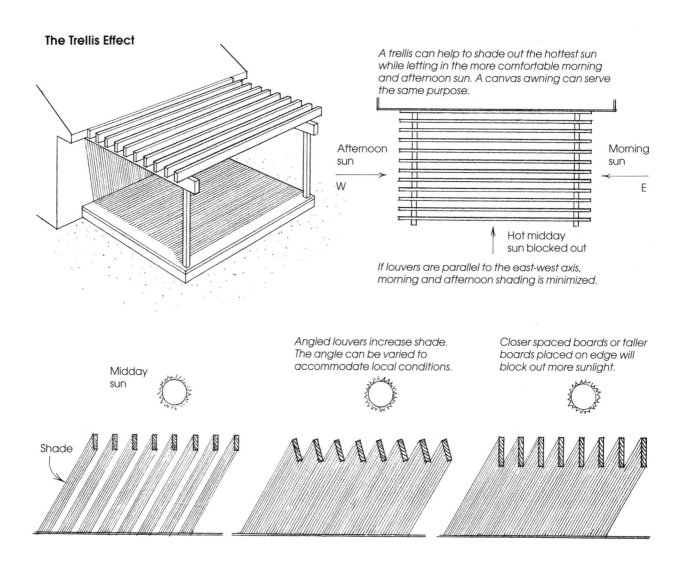

*A trellis can help to shade out the hottest sun while letting in the more comfortable morning and afternoon sun. A canvas awning can serve the same purpose.*

Afternoon sun

W

Morning sun

E

↑ Hot midday sun blocked out

*If louvers are parallel to the east-west axis, morning and afternoon shading is minimized.*

Midday sun

Shade

*Angled louvers increase shade. The angle can be varied to accommodate local conditions.*

*Closer spaced boards or taller boards placed on edge will block out more sunlight.*

highest and still allow plenty of sunlight earlier and later in the day. Greenery growing on the trellis can add to the shading.

Wind is a particularly capricious element that can complicate a deck design. If windy conditions are likely on a regular basis, you can take steps at the design stage to minimize the problem. Wind can be kept away much like sun, with fences and shrubs for example, or by locating the deck so that it's less affected by prevailing winds. On the other hand, if you want to catch mild breezes, don't include in the design any elements that would block them.

If you get a lot of rain during the warm months, you may want to design the deck so that it can be used while it's raining. This would entail covering at least part of it with some kind of overhead structure. Are you going to have to shovel the deck during winter? If

so, you need to consider where the snow will go and you'll want to look at railing designs that make snow removal easy.

**Privacy** You may need to make decisions about whether or not you want to see the neighbors and be seen by them. Do you want to be able to see the road, or would you rather be isolated as much as possible from traffic? If you want privacy, but the location of the deck doesn't allow it, you need to design a secluded nook.

**The view** A good view is an unquestionable asset to a deck, or to any piece of property for that matter. But a good view is often used carelessly. Views need to be handled with some discretion. If you design the deck so that it is fully and aggressively confronted with the best view, the view may become mundane, overused. The authors of *A Pattern Language* (Alexander et al, Oxford University Press, 1977) understand this well. Of a beautiful view, they say, "One wants to enjoy it and drink it in every day. But the more open it is, the more obvious, the more it shouts, the sooner it will fade. Gradually it will become...like the wallpaper; and the intensity of its beauty will no longer be accessible...." You must also keep in mind that the view from inside the house will likely change with the addition of a deck.

**Connections with the house**  It would be a mistake to choose the location of your deck based only on weather or other external factors. The relationship of the deck to the house is just as important. A location chosen because it provides a sunny play space for the kids won't get used if a parent can't easily see them from inside the house, especially from the kitchen. Likewise, a deck that is located to avoid inclement weather may as well not exist if it requires a long hike through the house to reach it.

The deck is an extension of the house. It can provide overflow capacity or just a pleasant place to feel the breeze. In either case, it needs access through a door. The door may be already installed, or it may have to be added. It may be from a bedroom onto a small, private deck, but most often the deck will serve more than one purpose and more than one or two people. Then the deck needs access from the more public areas of the home, such as the living room, dining room or kitchen. A lot of eating takes place on most decks, so access to the kitchen is usually a priority. The door to the deck should be a part of, and not a disruption of, the regular traffic patterns in the home. If possible, the door out to the deck shouldn't require a disruptive walk through the middle of an interior room.

Glass is a common choice for a deck door. It gives the deck a visible link with the inside. Sliding patio doors might suit some situations, but they tend to be energy inefficient. Atrium or French hinged doors with compression weatherstripping might be a more practical option (see *Fine Homebuilding* magazine, No. 68, pp. 42-46 for more on this subject). It's great if you have a door to the deck right where you want it. But often you won't. Adding a door to the kitchen can be a big deal because kitchens don't often have unused space available. A walkway or wrap-around deck connecting the main deck to an existing door may be easier than adding a door. If the deck leads to a pool or hot tub or the garden, it wouldn't be appropriate to have the access door in a carpeted area of the house.

A deck can drastically affect the amount of light that can reach some parts of the interior of the house. The deck we built for the Morris family completely shades the walkout basement below it, but this was the only way we could provide a full deck off of the main floor of the house. In this case, the clients understood the consequences and decided that they didn't much mind. If they had minded, however, because they

## Directing Traffic

*If you need to add a door to gain access to the deck, try to locate it so that traffic does not disrupt the use of a room.*

*Placing a door here directs traffic through the middle of this living room.*

*Placing the door here is better and allows both spaces to be used without disruption.*

**This casement window opens into a heavily traveled walkway on the deck, which is both inconvenient and potentially dangerous. The problem could have been avoided by installing a different type of window, designing the deck so that traffic would bypass this spot, or planning a different location for the window.**

hoped to turn the basement into living space some day, for instance, we would have had to revise the design. We might have moved part or even all of the deck away from the house, allowing access with a bridge or walkway. We might also have placed large openings in the decking at strategic areas above the basement windows, which would have allowed the sunlight to get through.

A deck can create some other unexpected problems. When we installed casement windows throughout the house, we didn't consider how they might affect the traffic flow on the yet-to-be-built deck. But, as can be seen in the photo above, casement (and awning) windows can be downright dangerous if they open onto a walkway. This situation could have been avoided by thinking ahead.

**Architectural considerations** How will the visual appearance of the deck affect its success? How can we best blend the functional and the aesthetic? A good place to start is with the ideas presented in the book *The Good House,* by Max Jacobson, Murray Silverstein and Barbara Winslow (The Taunton Press, 1990). Good design, they say, "is the production of harmony through the orchestration of strong contrasts." A good design is full of contrasts: variations in color, height, openness, light, solidity, enclosure and the amount of detail. Colors can be warm or cool. Spaces can be wide open or secluded. Railings can be bulky or thin to the point of near invisibility. Planned contrasts can actually become wonderful solutions to necessary compromises.

Just as the deck is a transitional space between indoors and out, parts of the deck serve as transitions to other parts, and these transitions offer opportunities for introducing contrast. Changes in level are one of the most common ways to emphasize a transition. A level change may be announced by a change in the railing detail, and you can accentuate the transition by adding planters at the top or bottom level.

Contrasts can be made with a change of materials or by manipulating the appearance of the same materials. A small stone walkway could connect two wooden decks. If only one material is used, it can be given contrast by installing it in a different direction or staining it a different color. Contrasts can be incorporated into all levels of the design. Large unrestricted areas can exist peacefully with smaller secluded spots. The rounded edges of a railing can offer contrast with the square corners of the posts.

Why is contrast good? Because differences sharpen our experiences. They also let us adjust our environment to our momentary needs. We can choose warm and sunny or dark and cool. We may want a maximum amount of social interaction or seclusion for contemplation. At the very least, decks should have small areas that are more isolated than others, such as a hidden bench in a strategic location, perhaps obscured by a level change or some shrubbery.

**Level changes** By changing levels on a deck, you can create the effect of two or more small decks, allowing each area to serve separate functions at any given moment. A level change can also allow the deck to follow the contours of the land more easily. This can prevent a house on a hillside from having a deck that towers way above ground level. The more a structure follows the natural topography of the land, the more it appears to belong there. Level changes provide edges and steps for seating. They can make the deck,

## Let the Sun Shine In

*A deck can block sunlight to lower level windows and doors. You can offset this problem by leaving an opening in the deck or by separating the deck from the house with a thin walkway.*

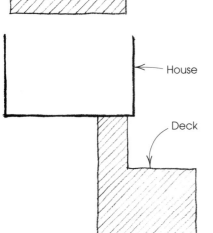

House

Deck

House

Deck

## Contrast in Railings

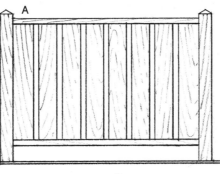

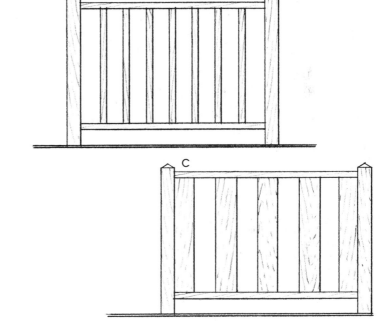

A

B

C

*Railings A and B have the same spatial relationships, only with reversed contrasts of open and closed. Either one presents greater interest and contrast than the monotonous regularity of Railing C.*

## Contrasting Openness and Seclusion

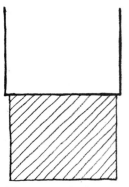

Wide-open deck

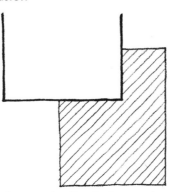

*By wrapping the deck around a corner, a little seclusion is added to the openness.*

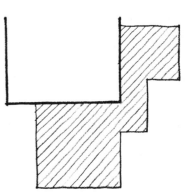

*Greater seclusion is created by separating the deck into two areas, connected by a walkway.*

Level changes in the deck add interest through contrast, allow for a gradual decline in the deck to conform with the sloping site and help maintain the view from inside the house.

### Level Decks on Sloped Sites

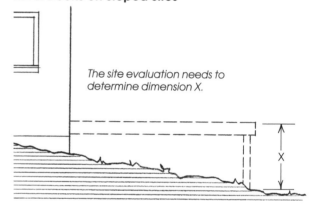

*The site evaluation needs to determine dimension X.*

and therefore the house, more visually appealing. And, where a large deck might obstruct the view from the inside of the house, a level change can lower the obstruction and thus improve the view.

**Complementing the house style**  Decks are a relatively recent architectural creation. Today, new houses are often designed with the deck in mind, but adding a deck to an old house can require some special consideration. For most older, traditional house styles, this probably means that a deck should not be designed to be too prominent or avant-garde. It might be best located in the back of the house, with a lower and more courtyard-like profile. On a Victorian house, the deck railing might mimic the ornate trim on the house. For an older brick house, brick could be used for the deck's foundation, as railing posts or in a connecting walkway. A freestanding deck, separated by some distance from the house, could maintain the stylistic integrity of the house. A new deck on an old house can also open up a variety of challenging contrasts. Some parts of the deck, such as the railings, can be painted in a manner that complements the paint on the house. Finally, on some traditional houses, a deck would be nothing but an eyesore.

A deck can soften the height of a tall house by being designed with a strong horizontal presence. Similarly, a long ranch house can be broken visually by locating a deck near the middle, while putting the deck on the end of the house would greatly exaggerate its flatness.

## Site Evaluation

The designer must know the site before beginning the design. A proper site evaluation will usually result in a site drawing, which in turn will guide the design of the deck. The site drawing will locate any pre-existing conditions that may influence the shape, look and location of the deck. It will show house and driveway locations, property lines and neighboring house locations, gardens, large unmovable objects such as trees and utility poles, water and sewer lines, and compass directions.

The site evaluation needs to look at the topography. On a sloped site, the extent of the slope needs to be determined and noted. Steep slopes may allow for more level changes and require more stairs. They may also require greater structural support with taller and

## The Site Plan

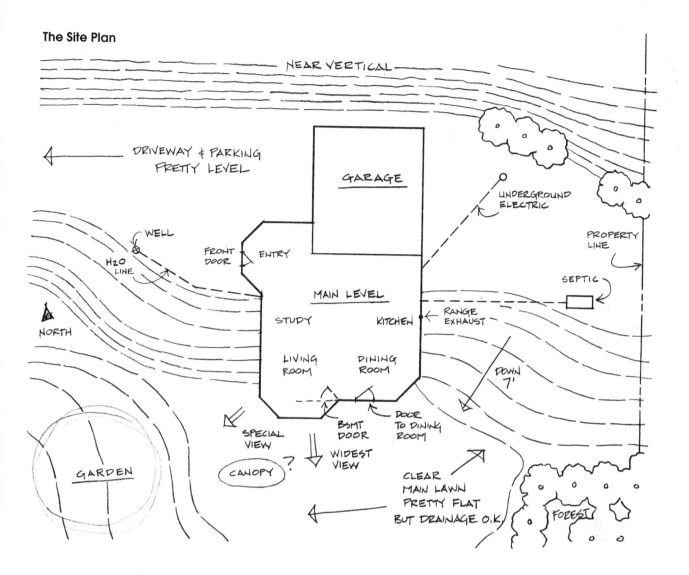

more frequent posts. This information is critical for designing a properly engineered structure.

To prepare a site plan, begin by placing some stakes around the proposed edges of the deck. Next, unless the site is flat, you need to find the difference in elevation between the highest and lowest ground points on the staked perimeter. With this information, you can determine the approximate length of the longest posts on the deck. You can measure this distance several different ways, using a carpenter's level, a line level, a water level or a transit.

The site plan should show the location of the house in bold dark lines, including the dimensions. Add the location of existing doors and note the use of rooms adjacent to the proposed deck. All buried hazards, such as utility lines and tanks, are usually represented by dashed lines. Changes in the ground elevations are marked with contour lines, as on topographical maps. The closer together they are, the steeper the slope, but writing in the total drop is easier than worrying about using the correct number of lines. If you've staked the approximate location for the deck, be sure to note the expected height above grade at the corners. Finally, add any trees and shrubs by shading or drawing circles to show their position.

Staking the deck at this point gives you a good chance to start visualizing the final product. Look for potential obstructions or easily overlooked conveniences like hose bibs and electrical outlets. Note these on the site plan and begin trying to figure out what you want to do about them. You may want to get a plumber or electrician involved in the project at this point.

The site plan needs to indicate where all utilities come into the property. Overhead electrical and phone lines and underground electrical, water, sewer and gas lines have to be located. You may want to think twice before building a deck over your septic tank. If you aren't sure about underground lines, check with the utility companies. Rerouting any utility will be a serious expense. You can usually plan the footings so that they don't affect underground lines, but think about what would happen if a broken line had to be dug up some day.

And don't ignore the overhead lines. I once framed an entire deck before realizing that I could now reach out and grab a 240-volt service entrance line. Building codes and common sense won't allow this kind of error to go uncorrected. Our only viable option in this situation was to raise the line. This was an expensive conclusion to a problem that could have been avoided had we just looked up.

## Sizing the Deck

The size of the deck you build will obviously depend a lot on your budget, but there are other factors that should influence the size, though I must admit that these can be very subjective. I consider 8 ft. to be the minimum dimension for any functional part of a main deck area. For transitional areas such as walkways, 6 ft. will work, and 4 ft. would be an acceptable minimum for a short passageway. I like to keep exterior stairs 4 ft. wide if possible, though you can get by with 3 ft. on little-used stairs. Whenever the deck stretches to a dimension that requires joists longer than 16 ft., it's likely to need extra beams, which may double the amount of foundation work required. But note that if the 16-ft. dimension runs perpendicular to the joists, this extra work won't be necessary because the length of the joists isn't affected. An 8-ft. by 12-ft. deck would be minimally adequate for four sedate people, but would rapidly become crowded with the addition of a dog, picnic table and barbecue grill.

The size of the deck must also be in scale with the size of the house. A small house will be better served with several small decks, separate or on different levels, than with one large deck. A deck on a large house doesn't look right unless it has spacious unbroken areas. Francis Ching, in *Architecture: Form, Space and Order* (Van Nostrand Reinhold, 1979), offers many suggestions for choosing balanced proportions and scale in building.

## Estimating Costs

Building a deck is a relatively easy job for a motivated do-it-yourselfer with basic tool skills. But the decision of whether to build it yourself or hire someone else to do it may depend on price. There are no absolutes about costs in construction projects. The price of materials, especially lumber, seems to fluctuate like the weather. And the cost of hired labor can vary seasonally as well as with the ebbs and flows of the local economy. In my area, a well built cedar (#2 or better) deck with pressure-treated joists will run at least $12 to $15 per square foot, but don't be surprised if conditions in your area are quite a bit different. This cost covers foundation through simple railings, labor and materials. Level changes, a set of stairs, some built-in seating, elaborate railings and a few coats of finish may well cost more. You can also expect to pay more, maybe a lot more, for curves, trellises, elaborate built-ins, clear lumber, landscaping, site problems or other unusual or exotic features. If you want a hot tub with the deck, you'll have to pay for the tub and probably a plumber and electrician as well.

Many do-it-yourselfers harbor the idea that professional builders make an extra killing on jobs by getting fat discounts on the price of materials, which they don't pass on the their clients. Don't believe it. After 15 years in the business in the same area, I still don't know how some lumberyards figure out who gets a discount. They all seem to have a different method. Volume is a major consideration, but sometimes the volume of materials on a single deck isn't big enough to earn a discount. If you can buy all of your deck materials at one place and at one time you're more likely to get a better deal than if you buy framing lumber, decking, finishes and hardware at different places. By "better deal" I'm talking about a discount of a few percentage points—not necessarily a big deal for a do-it-yourself job. Also, by frequenting one lumberyard and becoming a familiar face, you're more likely to get better pricing.

For most builders and do-it-yourselfers, the best way to get the lowest price is simply to make up a materials list for the whole job, being very specific about types, grades and lengths of lumber, and then give the list to several suppliers and ask for a quote.

If you're a do-it-yourselfer, you may save money on labor, but you won't save time. The DIY adage says figure out how much time you think it will take you, then double the figure. And if you're taking time off from a $30-an-hour job to build a deck, you won't even be saving money in labor. But deck building can be fun—it's a great way to learn some basic carpentry skills and build up confidence in your technical skills.

If you hire a builder or contractor, you can expect to pay him or her a percentage over labor and material costs to cover overhead and profit. This extra percentage may be insignificant with a builder who is satisfied with the hourly wage he or she charges or if times are tough. It can also be as high as 30% for a busy and well-established builder.

You can hire a builder who will do the job for a fixed price. The fixed-price bidder will have attempted to imagine the worst-case scenario and charged accordingly while still trying to remain competitive. This method requires the clients to be very specific about their requirements and to expect to pay for any changes made after the work has begun.

Many small but highly reputable contractors prefer to avoid fixed-price bidding by working on a "cost-plus" basis. This means that the client will be billed for the cost of all materials and labor plus a predetermined fee based on a percentage of the cost of the deck. This method removes some of the risk for the builder. I prefer to make the fee a fixed amount, which would change only if the client made major changes along the way. I believe that on the average, a cost-plus-fixed-fee job won't be any higher than a fixed bid, but that it will result in higher-quality work and greater satisfaction for both parties. If I were hiring a contractor on a cost-plus basis, however, I would want to check his or her references with a number of previous clients.

There are books available that claim to be reliable guides to costs for building projects. I don't know any builders who use these books, although I can imagine a newcomer finding some security in their pages. One of the problems with the books is that they try to flatten out a cost scale that, in the real world, is anything but flat. They work from average costs for materials and labor that are often inappropriate. In looking at the deck projects listed in *Home Improvement Costs for Exterior Projects* (R.S. Means, 1991) I found that the square footage prices were all higher than I charge, ranging from $16 to $30 per sq. ft. And I didn't find that the higher-priced deck necessarily matched my idea of greater complexity. I suspect that these figures assume a hefty percentage for overhead and profit. If you're a builder the only reliable source for estimating jobs is your own database compiled from past experience. If you're a home owner hiring a contractor, it's likely that prevailing economic conditions will have a large influence on the price of your deck.

Here's how the costs on a couple of hypothetical deck projects might break down. The first deck is 8 ft. by 12 ft. (96 sq. ft.), with pressure-treated joists and cedar decking. It has concrete piers and short posts, but because it's close to the ground it doesn't require stairs or a railing. Based on what I currently pay, the total cost of materials would be about $660: $215 for the decking, $180 for joists, $50 for a beam, $20 for posts, $40 for concrete and tube forms and $155 for hardware and finish.

Labor on this job would run about $800, figuring two workers taking two days. The total labor and materials fee would therefore be $1,460. If you as a builder charged $12 per sq. ft. for this deck, your fee would be $1,152, which is another way of saying that you would lose your shirt. Even at $15 per sq. ft. you would only make enough ($1,440) to cover labor and materials, with no margin for error and nothing left over to cover overhead and profit.

For this reason, small decks will often cost more per square foot than a larger, equally simple deck. And large decks with lots of extras will also cost more. The most economical deck (per sq. ft.) is 300 sq. ft. to 700 sq. ft. with a minimum of extras. Unfortunately, this type of deck will also be the least interesting.

The second hypothetical project is a 900-sq.-ft. deck with one or two stairs, two or three level changes, an interesting railing and all-around nice design. At $15 per sq. ft., this deck will cost $13,500. That sounds like a lot of money, but let's look at how it breaks down. Materials will run about $6,500, or about one-half the total. Labor would run about $6,000 (for 240 hours). This leaves about $1,000, which represents a 7% charge to cover overhead and profit. Note that in this example, the home owner would get a much nicer deck for the same price per square foot as the simpler deck, and the builder would make a modest profit. Everyone comes out ahead if you have the budget.

## Codes and Inspections

Most likely you'll be required to have your deck project approved by the local building department, and this will require compliance with local codes. It may also require inspections at various stages of the project, so talk with the folks at the building department before you get started and find out what they want. They can be an important source of information.

The biggest concern of the building department is safety, so your plans need to be specific about the structural elements of the deck. If you're considering any wiring or plumbing for the deck, then these aspects will require separate plans and separate inspections. You may also face local covenants or subdivision ordinances that stipulate deck sizes, styles, locations

## Drawings for a Basic Deck

### The Plan View

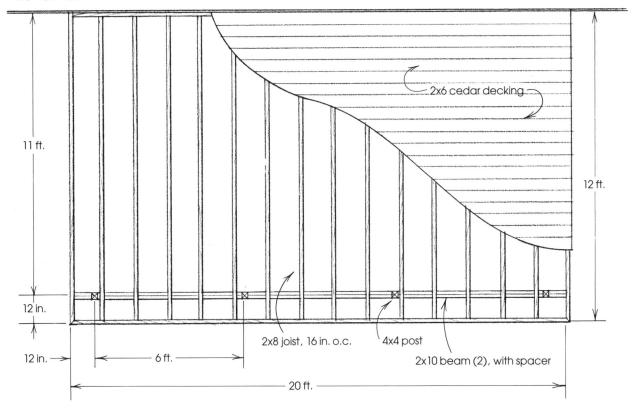

11 ft.

12 ft.

12 in.

12 in.

6 ft.

20 ft.

2x6 cedar decking

2x8 joist, 16 in. o.c.

4x4 post

2x10 beam (2), with spacer

### The Elevation Drawing

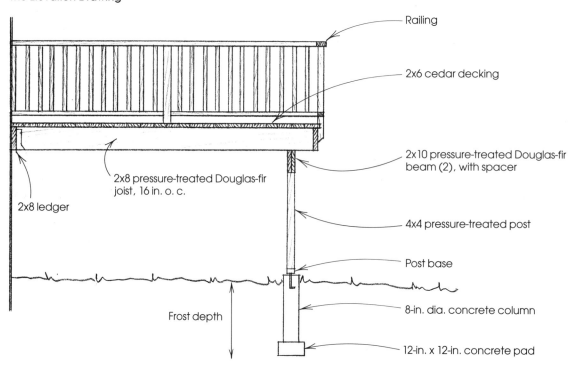

Railing

2x6 cedar decking

2x10 pressure-treated Douglas-fir beam (2), with spacer

2x8 pressure-treated Douglas-fir joist, 16 in. o. c.

4x4 pressure-treated post

2x8 ledger

Post base

Frost depth

8-in. dia. concrete column

12-in. x 12-in. concrete pad

and effects on your neighbors' solar shading. Setbacks from lot boundaries are another standard limitation.

Most of the code requirements have to do with how the deck is built. Some of the common requirements are:

1. The deck must be built to withstand the same structural load as the floors in the house, which is usually 40 lb. per sq. ft. (psf). In high snow-load areas, the deck will be loaded like a roof and the requirements may be 60 psf or more. Areas with high wind will require special fastening with metal straps. Earthquake-prone areas may require special fastening and additional bracing.

2. Railings are usually required if the deck is 30 in. above the ground. Railing heights need to be a minimum of 36 in. above floor level, and some areas may require 42 in. Balusters and all the parts of the railing need to be spaced so that there is no place where a 6-in. diameter ball (4 in. in many areas) can fit through. Even if the deck is close enough to the ground not to require a railing, it's a good idea to define the edges with posts, benches, planters or some other device. When both the deck and ground are covered with snow, it can be very difficult to see the edge.

3. Stairs usually need to be at least 36 in. wide (that is, 36 in. of usable space between posts). Handrails are required if the stair is more than four risers high, and handrails may be required on both sides if the stairs are more than 44 in. wide and more than 30 in. high.

4. Footings will usually need to extend down to frost depth, which is 48 in. where I live in Alaska.

## Wheelchair Accessibility

Building a deck that is accessible to someone in a wheelchair is generally required only on commercial and multifamily buildings, but I think it's a good idea for most residences. Especially if the deck serves the principal entrance to the house, a ramp would certainly be appreciated by a relative or friend who uses a wheelchair. Ramps can also make it easier to roll heavy items in and out of the house. To allow a wheelchair to get through, all passages should be at least 36 in. wide (48 in. would be ideal). The slope of the ramp should be not more than 1-in-12, and landings should be at least as wide as the ramp and 5 ft. long.

## Drafting

The final step in the design is translating all of the thoughts and ideas and rough sketches into plan drawings (see the drawing on the facing page). This can be a fairly simple chore for a basic deck, although it becomes considerably more difficult (and more important) on a complex deck design.

The basic tools for drawing plans are paper, pencil, ruler and a small drafting square. If you need to draw a lot of parallel lines, then a small drafting board and T-square will save you much aggravation. A 12-in. architectural scale will be a real time saver. Graph paper can be used instead of the square.

You'll need to draw the plans to a scale (¼ in. to 1 ft. is a good choice) that allows you to get the whole drawing on one sheet of paper, which should be at least 12 in. by 16 in. The first drawing you should make is a "plan view." This is the view from overhead, which should include the outline of the deck, the wall of the house, and the locations of joists, beams, posts and footings. Then you should draw enlarged details of special features, such as corner joints and miters, where you want to be sure you buy and use long enough boards. You'll also want detail drawings showing where boards lap over other boards, such as joists over beams, spacings between boards, locations of stairs and level changes, and other such items.

Then you need to draw a side view of the deck, called an "elevation." This view may show vertical distances, such as railing heights, posts and beams and level changes. You may want to draw side and front elevations. If the drawings are thorough and have been drawn to scale, they should be all that a competent builder needs to build the deck you designed.

Designing a deck, as I've tried to point out, is an evolving process. You may find yourself making changes right up to the last minute, and some even later than that. Don't panic. Some folks can handle tools but have trouble visualizing what it is that they're building. Others can tell you in advance exactly what something should look like, but are unable to use a screwdriver. All skills come with practice, and designing is no different. Observe, imitate and personalize, and eventually you'll find the deck that suits your needs. Decks are forgiving. You'll have to work hard to design a really bad one, but if you follow the information I've provided, designing a beautiful deck will be worth the effort.

# Building Materials

## Chapter 2

A well-designed deck is only as good as the paper it's drawn on until it has been constructed. And just as a good deck requires a good design, it also requires well-chosen materials. Wood must be selected both for its appearance and for its ability to stand up to the elements. The fasteners need to last as long as the wood. Choosing the right concrete and caulk requires some careful deliberations.

Material selection involves making up lists that specify the variety, sizes and quantities of items needed. This means that the materials can't be chosen until the deck has been designed from top to bottom. Buy the best materials you can afford, and then do the best job you can putting them all together. That way, the end product will be worth the investment.

## Wood

The main ingredient of a deck is wood—warm to the touch and beautiful to the eye, but hard to keep that way when used on a deck. Water, sunlight and negligence are a deck's enemies. Since it's not practical to keep a deck covered, it's important to choose woods that can endure tough conditions. It's equally important that the deck be designed and constructed to minimize water traps and maximize ventilation, for no matter how expensive or well-intentioned the choice, the wood's performance will be compromised if it can't get dry. The information in this chapter will help you choose the right wood for your deck, in terms of appearance, durability, availability and cost.

First and foremost, wood used on decks must be rot resistant. Most decks stay wet for extended periods of time, and woods that are not resistant can quickly decay. Cedar and redwood are naturally rot resistant, and other woods can be made rot resistant through chemical treatment.

You also want to use wood is that doesn't warp readily. It's not as simple as just choosing the right species, because there are many other factors that also influence the stability of a piece of wood—moisture content at the time of construction, drying method, milling technique, the size of the tree from which it came, and so forth.

Proper construction techniques can also minimize twisting, cupping, bowing and splitting boards. Although wood can't move radically after it's been nailed down, dimensional stability is a concern as the lumber cycles between wet and dry periods. Wood expands as it absorbs moisture and shrinks as it dries out. A wood's ability to remain intact and flat during these cycles is important. Inherent stability becomes even more critical if the deck is not properly maintained.

The relative emphasis that you place on rot resistance and stability will depend on the environment that the deck will exist in. For example, if the deck will remain wet most of the time, it won't shrink, and so it won't be as prone to warping, although you may have more problems with decay. Pressure-treated wood may have a longer life expectancy against decay, but only if it's properly maintained. If the deck isn't maintained, redwood and cedar may well stand up better.

Often, however, the biggest considerations in wood selection involve availability and cost. In many parts of the country you may have only one or two wood choices, although special orders (at a higher cost) are almost always possible. In the East and Midwest,

## How Wood Warps

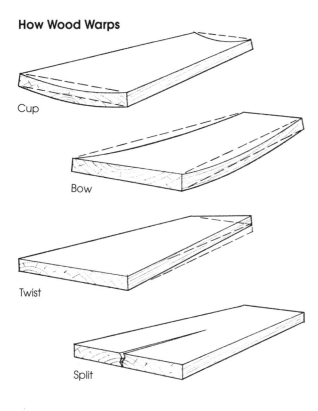

Cup

Bow

Twist

Split

pressure-treated Southern yellow pine is the most common deck wood, while redwood and cedar become increasingly available the farther you move west. And when considering cost, you need to weigh the difference between initial cost and life-cycle cost. Choosing a less expensive wood may cost you more in the long run in maintenance and repair.

Many decks are built with a combination of woods. For foundations and framing, it's common to use pressure-treated lumber, while the decking and railings are built with redwood or cedar. Even if you use the same wood for the whole deck, you can save money by using different grades for different purposes. On a pressure-treated wood deck, you might want to use a clear (knot-free) grade for the railing, #1 grade for the decking and #2 for the framing.

Choosing the right wood should also entail some environmental considerations. Do you want to contribute to the depletion of a rare species or the elimination of a stand of old-growth trees? Do you care about the environmental damage done in the harvesting of certain trees? How involved do you want to get with the use of insecticides and fungicides in the routine maintenance of your deck? You may find that answering these questions is a bit more difficult than you think. I don't intend to give you any lectures, but in

the course of this book I hope that you will at least feel reasonably competent to make good choices.

**Wet and dry wood** A lot of the wood used for decks is sold wet, with a moisture content (MC) of 30% to 40% or more. Wood that has been pressure treated with waterborne chemicals is saturated during the process and is frequently sold that way. The best quality pressure-treated lumber (and the most expensive) will normally have been kiln-dried both before and after treatment. Wet wood will shrink, which is easy to allow for, but it may also warp, which isn't. Even wood that has been kiln-dried to MC 19% often will shrink further after it's been installed, down to perhaps MC 5% in dry climates. It's fine to purchase wet wood for outdoor use, as it's less expensive and will produce an equally strong deck. But if you can dry the wood before installation, it will be lighter, easier to work with and less likely to split.

The best method for drying wet wood is to "sticker" the lumber as soon as you can after delivery. This involves stacking the wood with spacers between the layers so the air can circulate freely. Make sure to stack it in a well-drained area. The weight of the stack will keep most of the lumber from warping too badly as it dries, but you need to add some weight on top to keep the top layers from warping (concrete blocks or equally heavy objects work well). Cover the pile just enough to keep out rain and snow, but not so much that you trap moisture inside the pile. Old sheets of plywood work great—they shed water and also keep the lumber well ventilated and in the shade, out of the concentrated heat of the sun.

Wet wood that has been stickered for two weeks in warm, dry weather will have dried out a lot, but it could take months for the wood to approach the moisture content of kiln-dried stock. There is no absolute rule of thumb here, but it's safe to say that the longer the wood has had to acclimate to the surrounding environment, the better it will serve as building lumber. If you build with wet wood, you'll need to compensate for the shrinkage that will occur—gaps will widen between decking boards and beams may shrink, causing the structure to settle.

If you're planning any fancy joinery on the deck, don't even consider doing it with wet wood. Glued-up laminations for curved rims or handrails or tight-fitting miter joints just won't last if they're made with wet wood.

So how do you know if the wood is dry? It may look dry on the surface and still have a high moisture content. Wood that was kiln-dried to MC 19% a few

## Which Woods to Use Where

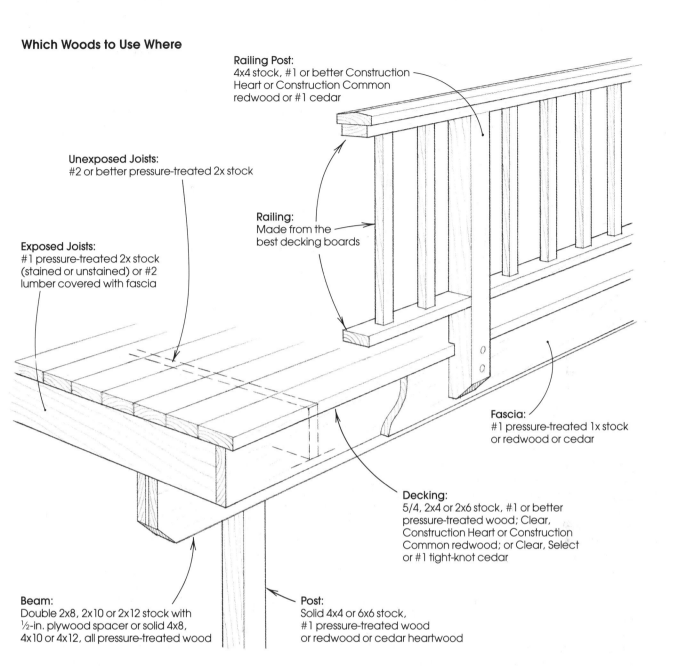

**Railing Post:**
4x4 stock, #1 or better Construction
Heart or Construction Common
redwood or #1 cedar

**Unexposed Joists:**
#2 or better pressure-treated 2x stock

**Railing:**
Made from the
best decking boards

**Exposed Joists:**
#1 pressure-treated 2x stock
(stained or unstained) or #2
lumber covered with fascia

**Fascia:**
#1 pressure-treated 1x stock
or redwood or cedar

**Decking:**
5/4, 2x4 or 2x6 stock, #1 or better
pressure-treated wood; Clear,
Construction Heart or Construction
Common redwood; or Clear, Select
or #1 tight-knot cedar

**Beam:**
Double 2x8, 2x10 or 2x12 stock with
½-in. plywood spacer or solid 4x8,
4x10 or 4x12, all pressure-treated wood

**Post:**
Solid 4x4 or 6x6 stock,
#1 pressure-treated wood
or redwood or cedar heartwood

months ago may have taken on some moisture in the meantime. Your lumberyard should have a moisture meter, which can give a quick reading of the MC. If you can't get your hands on a moisture meter, you can still tell a lot about the MC by crosscutting the middle of a board and examining the wood. Does it look or feel wet? Is there any evidence of water when you drive a nail into it?

Wood shrinks in all directions as it dries, but it shrinks much more in width and thickness than in length. "Plainsawn" or "flatsawn" boards, which describes the milling that characterizes most framing lumber, are more subject to large dimensional changes than "vertical-grain" (VG) boards. VG boards sell at a premium, and they are particularly useful for stair treads and other details requiring stability. VG boards also have a more durable face. You may find some VG boards in a normal delivery of framing lumber—if so, set them aside for special applications.

## How Wood Shrinks

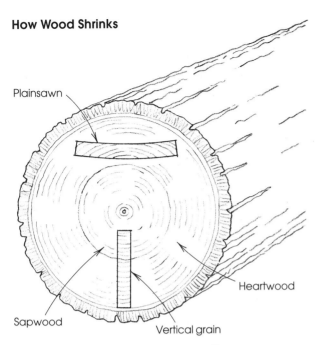

Plainsawn

Sapwood

Vertical grain

Heartwood

*A plainsawn board shrinks in width more than a vertical-grain board, and plainsawn boards are much more likely to warp.*

**Redwood** Redwood is a softwood that grows along the western coast of North America and is considered by many builders to have the most favorable natural characteristics for deck building of any indigenous wood. Its earthy, reddish-brown heartwood gives the redwood tree its name. As is the case with most naturally rot-resistant and insect-resistant species, the heartwood is the only part of the tree that has those beneficial characteristics. The creamy-colored sapwood contains none of the extracts and resins that give the heart its color and rot resistance. Some recent studies indicate that the increasingly available smaller, second-growth redwood (and cedar) trees may not have the level of decay resistance of the old-growth trees. This would argue for the use of treated wood for moisture-prone areas, especially if the wood is going to be buried. The reddish brown color of redwood can be maintained by regular applications of an appropriate finish (see Chapter 8). If left unfinished, the wood will initially become blackish, but eventually will turn to a beautiful silver grey.

Redwood is a very stable wood. It shrinks only about 5% when it passes from unseasoned to oven dry, less than almost any other wood, so splitting is minimized. The old-growth trees that yielded large amounts of straight-grained lumber helped give redwood the rep-utation for being warp free. Its natural stability also makes redwood less likely to cup than treated soft-woods. And even after extended weathering, redwood isn't prone to producing annoying splinters, which is a common problem with some other softwoods, such as Douglas-fir. Finally, because it is so soft, driving nails into redwood is a breeze.

There are some disadvantages to using redwood. In most of the country, it's about the most expensive op-tion for a deck. It also carries with it the stigma of en-vironmental problems, such as being harvested from clear-cut forests (see the sidebar on pp. 30-31).

Redwood is soft, and it needs to be protected against gouges from heavy objects. But it is adequate for most residential traffic areas. It's also brittle, especially when dry. When driving nails into the end of a board, espe-cially if it has been kiln-dried, it's usually necessary to predrill the holes to avoid splitting.

Redwood sawdust can be very irritating to some peo-ple. I have often experienced flu-like symptoms for a day or so after cutting redwood without using a dust mask. Unseasoned redwood will produce less of the ir-ritating dust, but you'll still want to wear a good dust mask when sawing. Redwood shavings don't seem to cause me any problem, probably because I'm not breathing them in.

Although I prefer to see the use of redwood reserved for decking and other areas where its beauty is dis-played, it can be sized to serve as joists, posts and other structural applications. In span tables, redwood is usu-ally lumped together with softwoods such as cedar and spruce. The California Redwood Association (see the Resource Guide on p. 150) can supply you with span tables that apply specifically to redwood.

The California Redwood Association also has a set of rules for grading redwood. They start with two main classifications: Garden and Architectural. The Garden grades cost less because they contain more knots. Both grades are subdivided into categories based upon the amount of sapwood the wood contains. The Garden grades that are most often used for deck building are Construction Heart, which contains no sapwood, and Construction Common, which contains some sap-wood. Two lower Garden grades, Merchantable and Merchantable Heart, contain more defects, such as larger knots, and are not recommended for decking. The Architectural grades would be used only on deluxe decks or for some special joinery details. Clear All Heart is the best grade available and contains no grad-ing defects. On the West Coast, redwood is widely available in various sizes and grades, kiln-dried or un-seasoned. In the rest of the country the selection may be much more limited, although you can look into special ordering.

**Cedar**  There are many varieties of cedar, but the most popular species for decking and siding is Western red cedar. It shares many of the favorable characteristics of redwood, including rot-resistant heartwood. Unseasoned cedar has a light brown color, but it will grey if left untreated. Cedar is dimensionally very stable, expanding and shrinking relatively little as it is exposed to the weather. Cedar does not produce nasty splinters as it ages. Cedar is not as strong as redwood or many other softwoods, and so it isn't used often for structural members such as joists. Although it is one of the softest woods used for decking, it will hold up to most residential uses.

I've built decks out of many different woods, but cedar is my favorite. Its heartwood of variegated browns and sapwood of a slightly different color don't overwhelm a house with lighter colored siding. Cedar has a pleasant smell that can give an exotic fragrance to the job site, although, like redwood, the sawdust can be irritating. The price and availability of cedar vary from region to region. In most parts of the country it will be easier to find than redwood, and it will be priced between pressure-treated pine and redwood.

Cedar is graded similarly to other Western woods, such as Douglas-fir, hemlock and spruce, but its use is generally limited to decking and siding (we're most often concerned with the appearance of the cedar than its structural strength). Cedar does have some unique specialty grades called Clear All Heart A and B, which distinguish boards with all heartwood and varying degrees of knots or other defects. These are the most expensive, but they provide the best assurance against rot. A step down in expense is the Select Tight Knot grade, which might also be called #1. This grade has few defects and some sapwood, but the boards are generally of a high quality. This is the grade I use most often. The #2 grade has more knots and more sapwood, and should be considered as a bare-minimum grade for use on a deck. Defects and sapwood can be cut out, but this can take a lot of time and produce a lot of waste. I like the modest contrast that a little sapwood provides to the heartwood and I've never experienced any rot problems by using it. But in a wet climate you would want to keep sapwood use to the absolute minimum. Full 1-in. cedar deck boards with radiused edges are sometimes available in a "patio" grade, which is similar to #2 or better.

These grading rules apply mostly to the nominal 2-in. and thicker material used for decking. Thinner stock, which might be used on a deck for trim work, is graded differently. Stick with the better grades if your budget allows. You can run into an occasional colloquialism in wood grading that will confuse you (and was probably intended to). I once saw some cedar that was graded "deck," which turned out to be a totally useless term that someone used because the wood was so bad they figured no one would want to use it on the side of a house.

Most deck-framing lumber will be stamped with the grade and species, but sometimes it's not. Pressure-treatment coloration may mask the stamp. And good grades of cedar, redwood or pressure-treated wood for decking may not be labeled at all, presumably to keep the ugly grade stamp from defacing the expensive wood. If you don't know your supplier, the best way to ensure that you're getting the grade you pay for, short of studying the rules in a grade book, is to shop and compare.

The same lumber grades can vary in quality from batch to batch and from mill to mill. I always like to look at the wood before I buy it, and I would special order wood only if I knew and trusted the lumberyard that was placing the order.

Redwood and cedar are usually available with one roughsawn and one smooth face (designated S3S) or two smooth faces (S4S). The rough surface is nice if you want to add some texture or accent a part of your deck. I like to use the rough face for fascias that cover ugly framing and for risers on stairs.

**Pressure-treated wood**  The use of chemically treated wood on decks seems to be increasing as the supply of naturally rot-resistant wood dwindles and the price increases. Almost any wood can be treated with one of a variety of preservatives, but some woods accept the treatment better than others. The predominant type of pressure-treated wood in the East is Southern yellow pine, while in the West, fir, larch and hemlock are more common. All of these woods perform well and can be used on most parts of the deck.

It's noteworthy that pressure-treated wood can last substantially longer than redwood or cedar. The Forest Products Laboratory in Wisconsin found that treated Southern yellow pine in contact with the ground can have a useful lifespan of about 42 years—about double what you can expect from redwood and cedar (see the chart below left).

Chromated copper arsenate (CCA) is the chemical most commonly used in pressure treatment, although ammonical copper arsenate (ACA) and acid copper chromate (ACC) are used in some parts of the country and on some types of wood. The treatment process involves placing a load of lumber in a special chamber

## Grading Lumber

*Grading standards for softwoods such as Southern yellow pine and Douglas-fir are as follows:*

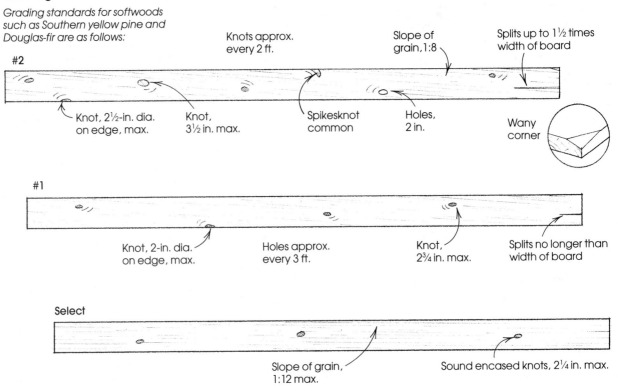

#2

Knots approx. every 2 ft.

Slope of grain, 1:8

Splits up to 1½ times width of board

Knot, 2½-in. dia. on edge, max.

Knot, 3½ in. max.

Spikesknot common

Holes, 2 in.

Wany corner

#1

Knot, 2-in. dia. on edge, max.

Holes approx. every 3 ft.

Knot, 2¾ in. max.

Splits no longer than width of board

Select

Slope of grain, 1:12 max.

Sound encased knots, 2¼ in. max.

| Life Expectancy of Wood in Contact With the Ground | |
|---|---|
| **Species** | **Years** |
| Southern yellow pine | |
| pressure treated | 42 |
| untreated | 3 to 12 |
| Western red cedar | 8 to 24 |
| Redwood | 20 |

*Source: Forest Products Laboratory*

and then forcing the waterborne chemicals into the cells of the wood by creating pressure or a vacuum in the chamber. The wood is sopping wet after the treatment, and sometimes it goes straight to the lumberyard that way. The wood can be kiln-dried after treatment, but that adds to its cost.

Pressure treatment does a much better job of impregnating the wood with preservative than you could do by dipping or brushing a preservative on the wood at the job site. It also saves you from having to handle these hazardous substances. CCA is toxic stuff, but if properly treated it won't leach out of the wood because the chemicals actually bond with the wood. Theoretically this means that no toxins should rub off on your skin when you're handling the lumber, but I prefer to play it safe. I wear gloves when I handle pressure-treated wood and avoid breathing in any sawdust. Treated wood should not be burned either, especially in a woodstove, because the fire will release the toxins into the air. I don't use pressure-treated wood anywhere it's likely to come in contact with food, such as on a tabletop, and I make a special effort to wash my hands before eating lunch if I've been handling treated wood. Some manufacturers of treated lumber may require you to brush some CCA on to any parts of the lumber that have been cut in order to keep the warranty valid, though I would try hard to avoid taking this step.

The amount of preservative absorbed by the wood will determine just how well preserved the wood is. The pounds of preservative absorbed per cubic foot of wood is called the retention level, and it should be stamped on every piece of treated lumber along with the stamp of the American Wood Preservers Association (AWPA). You can use wood with different retention levels for different purposes. On a deck, you could use 0.25 for above grade and 0.40 for ground contact. Many builders will use 0.40 for the whole deck, just to be on the safe side. (Your lumberyard may also carry 0.60, which is only needed on wood-foundation systems.)

CCA-treated lumber will have a green tint that will gradually turn to a grey over a year or two of exposure. I don't like the color even for that long, so I prefer to use prestained pressure-treated lumber for any exposed portions of a deck, even though I pay a premium for it. I try to get a stain that matches the cedar I use for decking and railings. There are other options to buying prestained treated lumber. One is to use unstained wood for the framing and then cover the exposed boards with a more attractive trim board, such as ¾-in. cedar. Or you could stain the wood yourself, but it must be thoroughly dry before you do.

Southern yellow pine is strong, stiff and hard, and it is more easily and successfully treated with CCA than other softwoods. However, Southern yellow pine has a reputation for being unruly, and it needs to be dried carefully to prevent warpage. It is also subject to shelling, a common defect in some softwoods in which the grain tends to separate after cycling between wet and dry over time, forming wicked splinters. Shelling can be controlled, however, by protecting the wood with a water repellent. Southern yellow pine is usually grouped with Douglas-fir on span tables, indicating similar structural strength.

Douglas-fir, which grows in the West, is a strong and heavy wood that tenaciously holds a nail and doesn't split or warp easily, which makes it a popular choice for framing. It is one of the most widely available treated woods in the West, but because it is a bit resistant to the standard pressure-treating process it often has to be incised before treatment to increase the penetration of the preservative. Douglas-fir is prone to shelling and should be treated regularly with a water repellent. Pressure-treated larch is available in some areas, and though I have no direct experience with it, it is graded for strength in the same class as Douglas-fir. Douglas-fir-south is a different and somewhat weaker variety of wood.

Hem-fir is another widely available classification of wood. It groups together several species of fir with hemlock, which is the predominant member of the group. Hem-fir is weaker than Douglas-fir and it's more prone to warping and splitting. But it's cheaper too, and is a common variety of treated lumber.

The process of pressure-treating wood does not change its grade or span rating. Boards are graded by the number of defects they contain. Structural grades for framing, which apply for use as decking as well, begin with Select as the best, followed by #1, #2 and #3. Usually, the framing lumber at the lumberyard will be graded #2 or better, and this should be adequate for concealed framing. However, if you're going to use treated lumber for the deck boards, you should use at least #1 boards, which will make for a much better looking deck. You can often find #1 grade 5/4x6-in. deck boards, with the edges already rounded.

**Other wood choices** There are other naturally rot-resistant and decay-resistant woods than redwood and cedar. They include cypress, locust, Osage orange and white oak. Of these, however, only cypress would be a suitable candidate for deck use. It's durable and soft enough to work easily, but it's prone to warping. Cypress may be available at a reasonable price in some parts of the southern U.S.

Regular, untreated framing lumber of any species should be considered a last choice under nearly all circumstances. For one thing, many codes won't allow its use for framing a deck. For another, any initial savings in the price of the lumber will be offset by a shorter (and quite likely a much shorter) lifespan. In dry cli-

**Keep a separate scrap pile of pressure-treated wood. It should be buried or taken to the dump. Burning it creates toxic smoke.**

mates, it is possible for a well-constructed and well-maintained deck built entirely out of untreated fir, pine, hem-fir or spruce to last a number of years, but untreated framing lumber is a risky choice.

Some people may be tempted to build a deck with regular framing lumber that they treat themselves with some less-toxic preservative. The authors of the book *Common-Sense Pest Control* (William Olkowski, Sheila Daar and Helga Olkowski; The Taunton Press, 1991) are authorities in finding "least-toxic solutions" for all kinds of problems. While warning against the use of pentachlorophenol ("penta"), arsenates and creosote, they provide a list of alternative preservatives for deck woods that includes copper and zinc naphthenate, TBTO compounds and borate preservatives, which are often available at paint stores. They recommend dipping the wood into the preservative for at least three minutes, although it could also be brushed on after construction. Some of the negative aspects of this method are that it exposes the user to the preservatives (which, again, are "less" toxic, not "non" toxic), requires user vigilance in the storage and disposal of the material, takes a lot of time and, finally, protects less than common pressure treatment of lumber.

One final wood choice is suggested by the builder and architect Paul Bierman-Lytle, who is an expert on environmentally responsible building. Writing in *Fine Homebuilding* magazine (No. 73, Spring 1992), he recommends the use of a hardwood called pau lope, which is grown on Brazilian plantations. He claims that the wood is dense, naturally resistant to insects and rot and that it requires no finish. Finally, he says, it is priced comparably to cedar and is actually cheaper than the better grades of redwood. I hesitate to recommend something I haven't tried myself, but I think it deserves looking into (see the Resource Guide on p. 150).

# Decks and the Environment

I feel some uncertainty about using or recommending redwood, cedar or other semi-speciality woods, especially when practical, albeit less beautiful, alternatives are available. Using these woods, indeed any woods, requires that we be reminded of the effects of the use and abuse of natural resources on the local and global environment. The issues involved can't be reduced to simple solutions. Those who say, for example, that jobs are more important than the environment, or vice versa, are missing the point and postponing meaningful answers. In this book I provide a lot of technical information and aesthetic considerations on choosing wood for a deck. In this section I would like to balance that approach with a few thoughts from an ecological point of view.

Many people just plain hate the idea of using wood that's been given a hi-tech injection of toxic preservatives (I'll confess that I'm one of them at times), so they opt for a more "natural" choice, say redwood or cedar. But the best redwood, the fabulously large, clear boards of beautiful and decay-resistant heartwood, comes from harvesting the largest old-growth trees. As a vital and irreplaceable part of our remnant forest ecosystem, the remaining 75,000 acres of ancient redwoods in the Sierras, I feel, should be left alone. Smaller, second-growth trees are a more sustainable source, with about 2.5 million acres of coastal redwood in production, and should be used instead (these numbers come from the Public Forestry Association). But this means using smaller boards with more knots and less valuable heartwood. Using salvaged redwood is a great idea, and there are a number of suppliers of this recycled lumber on the West Coast (see the Resource Guide on p. 150).

I used cedar for years thinking that it didn't have the same environmental problems as redwood, but I'm coming to understand that this view was naive and that cedar just hasn't received the same public scrutiny. Some organizations believe that it's more responsible to use redwood than cedar because redwood is regularly replanted after harvesting, while the cedar industry has yet to implement an adequate renewal system.

Ancient cedar forests deserve the same protection as the old-growth redwoods. And since cedar grows more slowly than redwood, it's perhaps less likely to compete as a

## Fasteners and Hardware

Choosing the right hardware and fasteners for a deck job is every bit as important as selecting the right wood. The most common types of hardware are joist hangers, post bases and caps, flashing and an assortment of anchors, clips and ties. You'll also need a variety of nails, screws and bolts. Just as you want wood and wood treatments that are intended for exterior use, you'll also want fasteners and hardware that are specially made to resist corrosion. With today's long-lasting pressure-treated wood decks, it's possible for the wood to outlive the nails.

**Nails** Nails come in a variety of sizes and types and with several coatings and treatments, so it is possible to select nails that provide the optimum corrosion protection, holding power and invisibility after they've been installed. Framing nails have a large flat head and are used to nail joists, beams, ledgers and other large members. Framing nails are available as either common nails or box nails. I prefer the thinner box nails for most work; they have less of a tendency to split the wood when nailed close to the end of a board, especially if the points are blunted first.

Sixteen-penny (16d) framing nails, which are 3½ in. long, are used for nailing rim joists to floor joists and 2x deck boards to the joists. Use 10d nails for 5/4 decking. You'll also need a good supply of 8d nails (2½ in. long), which are used for nailing ¾-in. material such as trim and plywood, for toenailing and for temporarily tacking pieces together.

For corrosion resistance in framing, I strongly recommend using hot-dipped, zinc-coated galvanized nails. "Hot galvies" are a dull grey color with a rough surface that keeps the nails from working loose over time. Stainless-steel nails are very resistant to corro-

---

tree-farming product. Tree farming also has some problems of its own—as with any crop, maintaining high yields on the same soils gets increasingly difficult over time as the soils are depleted, which can lead to the use of various chemicals to "replenish" the soil, and so on into the abyss.

Exotic tropical hardwoods occasionally find their way into deck building, but this is an inappropriate use of limited resources, and it's also expensive. The hastening destruction of the tropical rain forests has been linked to increased carbon dioxide levels, ozone depletion and our health. Of course, the people who live in these tropical countries need money to survive and so the solution is not simply a matter of no more cutting. One facet of the answer appears to lie with the proper recognition given to the value of all woods in the rain forest. By exploring the value of many lesser-known species of tropical timber the value of the whole forest is enhanced. This will lead to less waste and a broader base for the economy. Tree farms are

a source that may improve the supply of certain exotic woods without damaging undisturbed regions of existing forest. However, these farms are hard on the soil, produce lower-quality timber and are subject to the diseases of monoculture.

Unfortunately, some unusual tropical woods are often recommended simply because they are temporarily available, but eventually prove to have their own set of problems that come to light in time. When you add the complexities of politics and money, the solution for tropicals is intricate. Unless tropical woods are proven to be harvested in a socially and environmentally responsible manner, I suggest the conservative approach. For now, that means using them only when there is no substitute and carefully investigating a variety of information sources to ensure the correct choice.

So where does this leave the conscientious home owner or builder? My intention is not to discourage anyone from building and enjoying a

deck. Rather, I'm trying to make people aware of what lies behind their choices and to understand one of the central laws of ecology: "All things are interconnected." We all need to use materials conservatively and to use prime materials only where they are essential. Whenever possible, use local and recycled materials.

Maybe you would be happier building a stone or brick patio than a wood deck. Building smaller decks and homes would certainly be a step in the right direction, but that's a tough solution for a builder, whose income is based on the size of a client's deck or house. As a builder, home owner and parent, I often wonder how I should balance my need for income and comfort with my moral obligation to leave the world healthier than I found it. We all need to search for those balances, and even if the solutions seem beyond our individual reach, the biggest mistake would be to give up trying. For a list of organizations that are working on issues of responsible wood use and forestry, see the Resource Guide on p. 150.

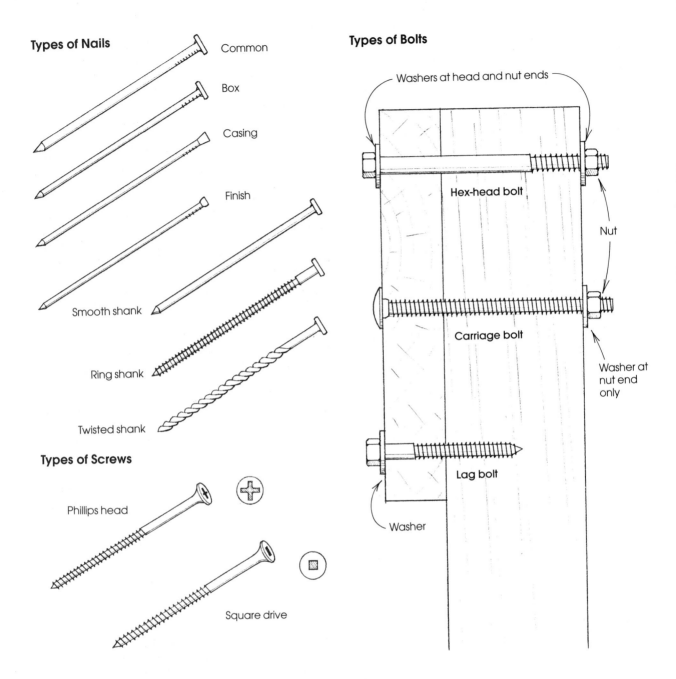

## Types of Nails

Common

Box

Casing

Finish

Smooth shank

Ring shank

Twisted shank

## Types of Screws

Phillips head

Square drive

## Types of Bolts

Washers at head and nut ends

Hex-head bolt

Nut

Carriage bolt

Washer at nut end only

Lag bolt

Washer

sion, but they cost a lot. They should be used for below-ground nailing and in marine environments.

For finish work, such as on railings and benches, you should use either finish or casing nails, which have smaller heads than framing nails. I prefer casing nails for exterior work because they hold a little better. I usually use framing nails for fascias, stair risers and other wide trim boards because they need the larger head to hold them in place.

For nailing down decking, you have more choices than for framing, and proper selection is more critical because of the visibility of the work. In addition to being corrosion resistant, deck nails need to have good holding power, and the less visible they are after installation the better. Ring-shank nails offer the best holding power, and twisted-shank nails will hold better than smooth nails. On the other hand, if you need to remove any nails, you'll appreciate smooth shanks.

When I hand nail redwood or cedar decking, I use a decking nail. This newer version of the casing nail is hot galvanized and has a slightly larger head than a casing nail and a twisted shank. In some climates (and perhaps with some manufacturers' nails), hot galvies react with cedar and redwood to produce a black discoloration. I haven't experienced this problem to any great extent, but if you want to play it safe, choose the ring shank, stainless-steel decking nails made by Swan Secure Products (see the Resource Guide on p. 150), which have small, flat heads and a slender shank that reduces splitting. These nails sell for about $6 to $8 a pound.

Special pressure-treated lumber nails, with a double coat of molten zinc, are available from W.H. Maze (see the Resource Guide). They cost less than stainless-steel nails and are available with smooth, twisted and ring shanks. Their large heads aren't very discreet, but they'll help keep the boards from working loose.

Stainless-steel and galvanized nails are also made for pneumatic nailers, which is how I usually do my nailing. For my Senco nailer, I use Senco twisted-shank nails called Weather-X. These electroplated nails have a double-thick coating that I find to be much better than the commonly available electroplated nails sold by Senco. Weather-X nails cost 25% more, but I consider them worth the extra expense.

**Screws** A variety of corrosion-resistant screws are also available for use on decks. Screws can offer a lot of advantages over nails for fastening decking—they provide much better holding power, they can be removed easily if boards need to be repaired or replaced, and their heads are often smaller than a nail head, which makes them less visible. But these advantages come at a price—the screws cost more than nails, and they can take a lot longer to install, especially if predrilling is required.

For decks, you should use self-tapping and case-hardened bugle-head screws or special decking screws with either a Phillips or square-drive head. They should be either hot-dipped galvanized or stainless steel. GrabberGard screws (see the Resource Guide) have a ceramic coating that is particularly effective against corrosion. Screws can be installed using a standard drill driver, a drywall screw gun or one of the specialized devices discussed in Chapter 5.

**Bolts** Bolts—hex head, carriage or lag—are required for framing connections where strength is critical. Bolts have much better withdrawal resistance and shear strength than nails, which are critical qualities for jobs like attaching a ledger to the house. Bolts offer insurance that an important connection won't come apart under strain, such as when someone leans against the railing posts of the deck or when the bracing is strained during an earthquake.

Lag bolts have threads like a wood screw, only deeper and coarser, and are used when you only have access to the head side of a bolt. Washers should always be used with lag bolts to keep the small heads from digging into the wood. Lags are used to fasten ledgers to the house framing and for attaching metal hardware to large beams.

Carriage bolts have a smooth rounded head that covers a short section of square shank. Washers should be used only on the nut end of a carriage bolt, but care must be taken not to overtighten them. Because of their attractive appearance, I use carriage bolts on railings, benches, built-up beams and other wood-to-wood connections that are likely to be visible.

The standard hex-head machine bolt is best for a really tight fit when you can get to both ends of the connection. They're usually needed on a deck only to secure heavy-duty post anchors or to fasten a particularly large beam.

I usually use electroplated bolts on my decks, and although some brands are unpredictably better than others, I have not experienced any unacceptable corrosion. However, I would recommend hot-dipped galvanized bolts in a marine environment if the bolt will be subject to constant dampness or if its location can't easily be inspected. With all bolts on a deck, it's a good idea to tighten them again after the wood has had a year to dry out.

**Masonry connectors** If you need to attach a ledger directly to a concrete wall or fasten a bracket to a concrete footing, there are several types of connectors for the job. Using a lag shield involves inserting a chunk of soft metal into a hole drilled into the concrete. A lag bolt is then screwed into the insert. Lag shields have limited holding power because of the softness of the metal gripping the bolt.

A better alternative is a wedge anchor or sleeve anchor, which is a threaded stud with a wedge on the bottom end that is placed in a snug hole drilled into the concrete. Threading a nut onto the stud forces the wedge to expand, producing a very strong connection. Epoxy anchors rely on an adhesive to hold the bolts. They are strong and reliable, though some have to cure overnight before they can be stressed.

Bolting to the thin walls of hollow concrete blocks is not easy, nor is the connection very strong. You can try one of the connectors mentioned above fastened to the grouted section of a block wall or into a web in

## Types of Masonry Fasteners

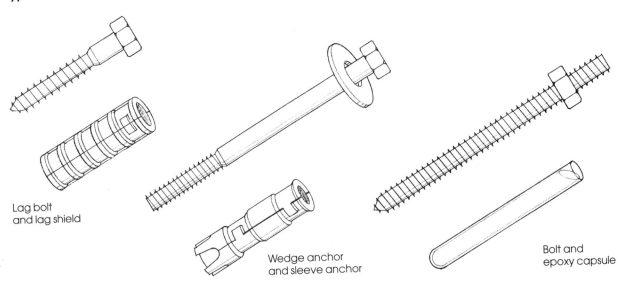

Lag bolt
and lag shield

Wedge anchor
and sleeve anchor

Bolt and
epoxy capsule

the block. If you must go into the hollow part of the block, there are special epoxy and mesh systems available for gluing in metal studs. Toggle bolts are sometimes used on block walls but they're only marginally effective and generally can't support heavy weights.

**Joist, beam and post connectors**  Various galvanized-steel connectors are used on a deck to secure posts to piers, beams to posts, beams to beams and joists to ledgers. Simpson Strong-Tie (see the Resource Guide) is one manufacturer that makes dozens of these connectors. They are available for handling double and triple joists, built-up beams, angled joists and extra-heavy loads. They aren't pretty, but metal connectors are a big improvement over toenailing.

Metal post caps and bases, which are used for connecting posts to timbers at the top or to concrete at the bottom, come in a variety of styles, strengths and sizes. The sturdiest bases can be cast directly into the footing, while others are attached with bolts or anchors after the concrete hardens. Most are made to fit 4x4s and 6x6s only, but some can be used for odd sizes as well. Hold-downs, straps, framing anchors and tie plates are used in a variety of framing connections, such as securing a joist to a beam, and these may be required by building codes in areas prone to earthquakes or hurricanes.

**Flashing**  Flashing is used to divert or block water. On decks, it's most commonly used to protect the connection between the house and the ledger. Some

builders like to flash along the tops of the joists to prevent water damage.

You can buy flashing in 8-ft. to 10-ft. lengths that have been bent to a variety of angles. Flat metal flashing also comes in 50-ft. rolls of various widths, but you can usually buy it by the foot at the lumberyard. Flat flashing is useful if you need to bend special shapes, but unless you have special tools you'll often have less than crisp corners and bends. You can have custom flashing made at a sheet-metal shop, but it will cost several times as much as stock sizes.

Flashing is available in aluminum, galvanized steel, stainless steel and terne. Aluminum flashing is much easier to bend by hand and will last forever if installed properly. However, it does move more under the influence of heat and cold and fasteners tend to loosen in it, eventually creating leaks. I prefer to use galvanized flashings. They are thicker and stronger and are less likely to suffer from mechanical damage over time. The zinc coating is chemically compatible with galvanized nails and so doesn't have the corrosion problems that can occur between dissimilar metals such as aluminum flashing and galvanized nails.

Another flashing product worth looking into is called Deck Seal (see the Resource Guide). It is made of a layer of aluminum foil with a flexible backing of sticky, rubberized asphalt, and it comes in rolls of 1½-in. to 6-in. widths. This flashing works best in applications where it is sandwiched between two pieces, such as on top of built-up joists to cover the joint between

## Joist, Beam and Post Connectors

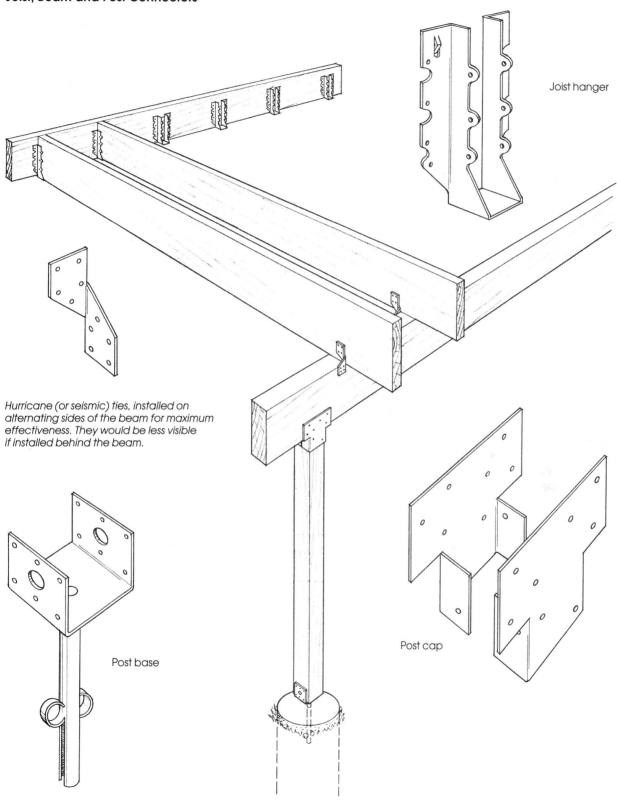

Joist hanger

*Hurricane (or seismic) ties, installed on alternating sides of the beam for maximum effectiveness. They would be less visible if installed behind the beam.*

Post base

Post cap

# A Lumber and Materials List

This list was prepared for the deck I built for the Morris family. A deck of a different size would require similar materials, but the quantities would change.

## FOUNDATION

| | |
|---|---|
| 40 | 1x4x8ft. Kiln-dried s4s Common pine |
| 9 | 8-in dia. x 12-ft. long fiberboard concrete tube form |
| 1 | yellow mason's line |
| 175 ft. | steel rebar, No.5 (5/8 in.), grade 60 domestic |
| 2 yd. | 2,500 psi, 4-in. slump concrete |
| 28 | 4x4 post base brackets |
| 50 | 1x2x24 wood stakes |
| 1 | 6 mil, 8ft. x 100ft. clear polyethylene sheeting |
| 1 lb. | 3-in. coated screws |
| 1 lb. | 1 5/8-in. drywall screws |
| 1 roll | orange flagging |

## FRAMING

| | |
|---|---|
| 5 | 2x12x16-ft. Kiln-dried, select, pressure-treated hem-fir |
| 15 | 2x6x8-ft. Kiln-dried, select, pressure-treated hem-fir |
| 52 | 2x8x16-ft. #2 and better, pressure-treated hem-fir |
| 24 | 2x10x16-ft. #2 and better, pressure-treated hem-fir |
| 26 | 4x4x8-ft. #2 and better, pressure-treated hem-fir |
| 3 | 2x12x8ft. #2 and better, pressure-treated hem-fir |
| 2 | 1/2-in. treated CDX plywood |
| 20 pr. | 4x4 post-and-beam connectors |
| 86 | 2x8 metal joist hangers |
| 15 lb. | 8d hot galvanized nails |

| | |
|---|---|
| 25 lb. | 16d hot galvanized nails |
| 5 lb. | 10d hot galvanized nails |
| 5 lb. | 1/4-in. joist hanger nails |
| 40 | 1/2-in. x 4-in. grade 2 lag screws, zinc plated |
| 49 | 1/2-in. plated washers |
| 90 | 1/4-in. x 3 1/2-in. grade 2 lag screws, zinc plated |
| 90 | 1/4-in. washers |
| 1 box | 3 1/4-in. galvanized pneumatic decking nails |
| 1 box | Duofast staples |
| 1 | 2x2x4-ft. galvanized angle iron |

## DECKING AND RAILING

| | |
|---|---|
| 165 | 2x6x16-ft. #1, s4s, green, tight knot cedar decking |
| 20 | 2x4x16-ft. #1, s4s, green, tight knot cedar decking |
| 23 | 4x4x8-ft. #1, green, cedar posts |
| 4 | 1x8x8 #1 Common, s3s, green cedar |
| 100 | 3/8 x 3 1/2 carriage bolts, plated |
| 100 | washers |
| 100 | nuts |
| 3 | 1/2-in. x 7-in. lag screws, plated |

## FINISHES

| | |
|---|---|
| 2 | 5 gal. natural wood finish, clear |
| 1 | gal. thinner |
| 2 | 9-in. roller covers |
| 2 | plastic tray liners |
| 2 pr. | disposable gloves |

them (some builders use felt paper for this purpose). It would also serve well to seal and flash water-absorbing end grain in places where maintenance is difficult, such as under treads on the carriage of a stairway or on top of posts that are carrying beams.

**Caulk** The spot most likely to need caulking on a deck is at the connection of the deck and the house, especially at the bottom and ends of the ledger and the ends of the flashing itself. You should also caulk around added hose bibs or thresholds. Silicone caulk is the all-around favorite. It is relatively cheap and it comes in a variety of colors. Its disadvantages are that it doesn't paint worth a hoot and it doesn't like to stick to oily woods like cedar. A good alternative is rubberized acrylic latex caulk, which is easier to "tool" and is a better choice for painting and for oily woods. It isn't quite as durable as silicone. Not recommended for use on decks are the older oil-based caulks, butyl rubber caulks and plain latex caulks. Polyurethane caulks, although the toughest of all the caulks, cost more and aren't usually necessary.

**Other materials** There are many other important materials that are needed on a deck job. For the foundation, you'll need to think about concrete, rebar, concrete forms and precast piers and pads. And after the deck has been constructed, it will need to be finished. These materials will be covered later—concrete in Chapter 3, and finishes in Chapter 8.

## Making a Materials List

After the deck has been designed, you need to make up a reasonably accurate lumber and materials list. This is useful both for estimating the cost of the job and for shopping around for the best price. It is best to break the list into categories that relate to specific suppliers or separate phases of construction—one list for concrete and foundation, one for framing and decking, one for railings and benches, one for hardware and fasteners and one for finishes. Your list won't be complete, but the more complete it is, the fewer trips back to the lumberyard you'll need to make. The list of materials shown on the facing page is for the 900-sq.-ft. deck I built for the Morrises. If you were building a more basic deck, your list might be about the same length, but the quantities would be proportionately less.

Before the lumberyard makes the delivery, give some thought to where you want the materials to be dropped. You want them close enough to minimize walking and carrying, but not so close that they'll interfere with the construction. Once your materials arrive on site, you're ready to start building the deck.

# A Gallery of Deck Designs

In the fall of 1991, the editors of *Fine Homebuilding* magazine asked their readers to send in photographs of decks they had designed and/or built for inclusion in this book. The overwhelming response provided ample proof that the once-humble deck has become a source of creative design and top-notch construction.

The following photos were chosen as much for their diversity as for their originality. Some of them portray full-sized decks; others are close-up shots of railing details or transitions between levels. A few of the decks demonstrate thoughtful integration with challenging sites or house designs, while others represent flights of fancy. Together, they provide a feast of inspiration for the deck builder and designer.

DESIGNER: Barb Sarapas and L. T. Swistoski
BUILDER: Opus Two Specialties
PHOTOGRAPHER: Helen Rickman
LOCATION: St. Paul, Minn.

DESIGNER/PHOTOGRAPHER: Daryl S. Rantis
BUILDER: Outdoor Rooms
LOCATION: Glenellyn, Ill.

DESIGNER/PHOTOGRAPHER: Cliff Chatel
BUILDER: Hammer & Nail Construction
LOCATION: Richmond Beach, Wash.

DESIGNER/PHOTOGRAPHER: Mark W. Benzell
BUILDER: Inside & Out Remodelers
LOCATION: Edina, Minn.

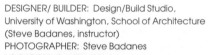

DESIGNER/PHOTOGRAPHER: G. Scott Geissler
BUILDER: G. Scott Geissler and Wayne Schneider
LOCATION: Moreno Valley, Calif.

DESIGNER/ BUILDER: Design/Build Studio,
University of Washington, School of Architecture
(Steve Badanes, instructor)
PHOTOGRAPHER: Steve Badanes
LOCATION: Seattle, Wash.

DESIGNER/BUILDER/PHOTOGRAPHER: Dennis Erdelac
LOCATION: Merrillville, Ind.

DESIGNER/BUILDER/PHOTOGRAPHER: Kevin Lane
LOCATION: Beaver Creek, Ore.

DESIGNER/PHOTOGRAPHER: Marc Maxwell
BUILDER: Golden Hammer, Inc.
LOCATION: Provincetown, Mass.

DESIGNER/PHOTOGRAPHER: Cliff Chatel
BUILDER: Hammer & Nail Construction
LOCATION: Bellevue, Wash.

DESIGNER/BUILDER/PHOTOGRAPHER: Kevin Lane
LOCATION: St. Louis, Mo.

DESIGNER/BUILDER: Walt Micks
PHOTOGRAPHER: Laura Micks
LOCATION: Huntington, Conn.

DESIGNER/BUILDER/PHOTOGRAPHER: Claude Audet
LOCATION: Magog, Quebec, Canada

DESIGNER: Ljubisa Jovasevic
BUILDER: Frank Massello
PHOTOGRAPHER: John Vecchiolla
LOCATION: Harrison, N.Y.

DESIGNER/PHOTOGRAPHER: Steve Badanes
BUILDER: Steve Badanes/Jersey Devil
LOCATION: Locust Grove, Va.

DESIGNER/PHOTOGRAPHER: Mark W. Benzell
BUILDER: Inside & Out Remodelers
LOCATION: Edina, Minn.

DESIGNER/BUILDER/PHOTOGRAPHER: Pat McCarty
LOCATION: Washington, Mo.
*This deck will be finished with wrought-iron railings and a spiral stair.*

DESIGNER/BUILDER: Brian Magyar
PHOTOGRAPHER: Richard Rosenhagen
LOCATION: Merrick, N.Y.

DESIGNER: David Durrant
BUILDER: Fine Line Woodworking
PHOTOGRAPHER: A. Samuel Laundon
LOCATION: Stow, Mass.

DESIGNER: William M. Taylor
BUILDER: Michael Gillooly and Doug Raleigh
PHOTOGRAPHER: J. Weiland
LOCATION: Weaverville, N.C.

DESIGNER: W. Michael Pachan
BUILDER/PHOTOGRAPHER: Michael W. Hunley
LOCATION: Cincinnati, Ohio

DESIGNER/BUILDER: D. L. Toman
PHOTOGRAPHER: Colleen Gallo
LOCATION: Lake Mohawk, Sparta, N.J.

DESIGNER/BUILDER/PHOTOGRAPHER: Walter Stevens
LOCATION: Barryville, N.Y.

DESIGNER: Barbara Thornburgh Carlton
BUILDER: DRP Construction, Randy Pain
PHOTOGRAPHER: Barry J. T. Carlton
LOCATION: San Diego, Calif.

DESIGNER/BUILDER/PHOTOGRAPHER: Edward Sprouts
LOCATION: Columbus, Ohio

DESIGNER: Fred R. Klein
BUILDER: Sound Construction Co. Inc.
PHOTOGRAPHER: Dean Photography
LOCATION: Eastsound, Wash.

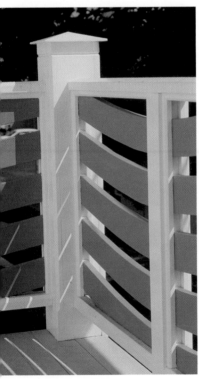

DESIGNER: Philip J. Bosko
BUILDER: Bob Goodman
PHOTOGRAPHER: Thomas Hahn
LOCATION: Mason's Island, Conn.

DESIGNER: Joe Bayless
BUILDER: Archadeck
PHOTOGRAPHER: Dwayne Snow
LOCATION: Charlottesville, Va.

DESIGNER/PHOTOGRAPHER: Arlene Barra
BUILDER: John Morris, Jr.
LOCATION: Brookfield Center, Conn.

DESIGNER: Harold Borkin
BUILDER: Richard Wineberg/Earthwood Builders
PHOTOGRAPHER: Susan Wineberg
LOCATION: Ann Arbor, Mich.

DESIGNER/PHOTOGRAPHER: Andrew Diem
BUILDER: Grays Building & Restoration
LOCATION: Chevy Chase, Md.

DESIGNER/BUILDER: Jonathan B. Ricker and Nick Kayfez
PHOTOGRAPHER: Jonathan B. Ricker
LOCATION: Mt. Clemens, Mich.

DESIGNER/BUILDER/PHOTOGRAPHER: Jonathan B. Ricker
LOCATION: Sterling Heights, Mich.
*"Screened deck" may be an oxymoron, but this Solartex screening blocks 75% of UV rays and all the bugs.*

DESIGNER/PHOTOGRAPHER: Andrew Smith
BUILDER: Upper Canada Deck Co.
LOCATION: Toronto, Ontario, Canada

DESIGNER/BUILDER/PHOTOGRAPHER: Edward R. Voytovich
LOCATION: Syracuse, N.Y.

DESIGNER/BUILDER: John Hemingway
PHOTOGRAPHER: Tom Rider
LOCATION: Los Altos, Calif.
*Courtesy: California Redwood Association*

DESIGNER/BUILDER: Mario Apap and Dean Virello
PHOTOGRAPHER: Mario Apap
LOCATION: Yonkers, N.Y.

DESIGNER: Joe Bayless
BUILDER: Archadeck
PHOTOGRAPHER: Joe Bayless
LOCATION: Macon, Ga.

DESIGNER: Fred R. Klein
BUILDER: Sound Construction Co. Inc.
PHOTOGRAPHER: Dean Photography
LOCATION: Eastsound, Wash.
*Clear acrylic sheets allow for a better view and screen the breezes better than conventional wood balusters.*

DESIGNER/BUILDER/PHOTOGRAPHER: Dave Weisler and Dick Mayberry
LOCATION: Lapeer, Mich.

DESIGNER: Helen A. Myers and Andy Reynolds
BUILDER/PHOTOGRAPHER: Andy Reynolds
LOCATION: Fairbanks, Alaska
*Most codes would require a railing for these stairs*

DESIGNER: Jolee Horne
BUILDER: O'Sullivan Landscaping
PHOTOGRAPHER: Steven B. O'Sullivan
LOCATION: Stanford, Calif.

DESIGNER/BUILDER: Dennis Whitworth
PHOTOGRAPHER: Doug Grainge
LOCATION: Philadelphia, Pa.

DESIGNER/BUILDER/PHOTOGRAPHER: Bruce Osen
LOCATION: Corvallis, Ore.

# Foundations

## Chapter 3

The humble foundation is a deck's connection with Mother Earth. Even though it takes a lot of work, building a good foundation will save a lot of maintenance headaches later on. A good foundation needs to be strong enough to support any expected load, and it must be sized to distribute this load adequately onto the existing soil. Additionally, the foundation must be installed so that it won't heave when the ground freezes, and it needs to be constructed of decay-resistant materials. Finally, the foundation needs to do all this at a cost that won't bust the budget. Your local building inspectors will want to know all about the foundation before you start building, and they should be a good source of information, too.

Unlike a typical house foundation, which spreads its load around the entire perimeter of the building, deck foundations are normally noncontinuous, that is, they concentrate loads at a series of isolated points. At each point, part of the total load from the deck is transferred through a beam and post above grade to a concrete pier below grade and on down to a buried, widened section called a footing. Alternatively, instead of a concrete pier, a partially buried, continuous wood post may run from the beam to the below-grade footing. At other times, the concrete pier may be so short that it is essentially indistinguishable from the footing.

When designing a foundation, you need to determine the number and size of the footings, what depth to bury them, what material to extend the footings to above grade and, finally, how to make the connections with the wood posts that support the deck. The foundation design is best done after the framing has been designed, because you need to know where beams will be located and how many posts will be needed. This chapter will help you make the right design decisions and then will explain how to lay out the foundation, dig the holes, form the piers and footings and deal with concrete.

## Sizing the Footings

The first step in foundation design is figuring out the footprint size of each footing, that is, the size of the area of contact between the bottom of the footing and the earth. Typical footprint sizes of footings for normal loads and good soil are 12 in. to 14 in. for square footings and 14-in. to 16-in. diameters for round footings. Local codes often dictate the minimum size of a footing. Although the information here is not meant to replace a professional engineer's design, it should give you a good guide for understanding and estimating the usually straightforward load and sizing requirements of a deck's foundation.

The first step in determining the minimum footprint size for each footing is to calculate what portion of the deck each footing supports. Then, multiplying the area supported at each footing by the expected load yields the load per footing. By dividing the load on each footing by the bearing capacity of the soil, you can determine the footprint size necessary for each footing. Changing any one of the variables in the calculation will affect the outcome.

Each footing supports a post, which in turn carries the load (transferred through the beam) that is applied on an area of the deck around each post and reaching halfway to the nearest neighboring post or ledger. To find the area in square feet for each footing, first multiply the length of the sides of the rectangle supported. Then multiply this area by the expected load per square foot to get the total load on each footing. Codes usually require that a deck be able to support a minimum of 40 lb. per square foot (psf) of "live load" (people and any movable items on the deck) and 10 psf of "dead load" (the weight of the permanent parts of the deck), for a total requirement of 50 psf (some codes require 60 psf). Although you won't use these numbers directly in calculating framing needs, as they're built into the span tables that you'll be using, they are useful in estimating the load on the footings. The larger the deck, the larger the expected load on the deck, which, assuming all other variables are held constant, will mean the need for larger footings.

Footings do not always carry equal loads. Corner posts may support less area and therefore have a smaller load than center posts, which could mean a smaller footprint size. Even so, it's often simplest and safest to make all footings the same size as that required for the largest load expected on any of the footings.

If the deck has multiple beams or small cantilevers, the footings may carry different percentages of the load, which will affect their size. And if the deck is going to support heavy objects, such as a hot tub or roof, or if the cantilevers get larger than 12 in., the normal load expectations won't be adequate. (Hot tubs usually rest on a separate slab, however, which does not affect the deck load.) If you have doubts or difficulties calculating the loads, you should consult an engineer.

## How Loads are Distributed

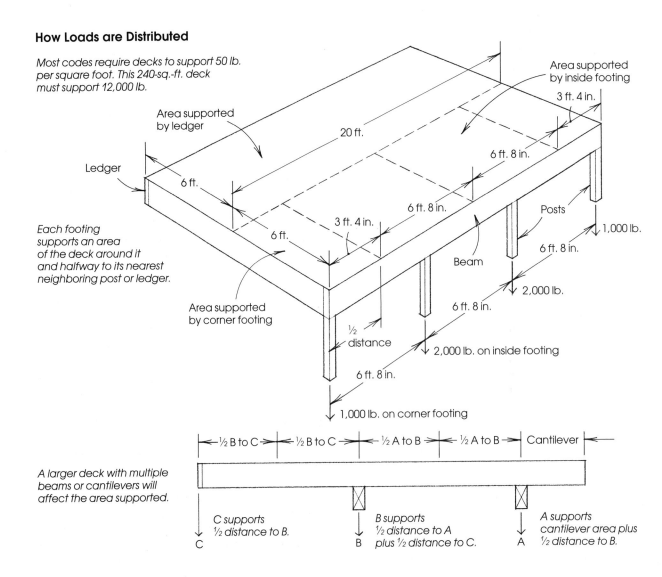

*Most codes require decks to support 50 lb. per square foot. This 240-sq.-ft. deck must support 12,000 lb.*

Area supported by ledger

Area supported by inside footing

3 ft. 4 in.

20 ft.

6 ft. 8 in.

Ledger

6 ft.

6 ft. 8 in.

*Each footing supports an area of the deck around it and halfway to its nearest neighboring post or ledger.*

3 ft. 4 in.

6 ft.

6 ft. 8 in.

Posts

1,000 lb.

6 ft. 8 in.

Beam

Area supported by corner footing

½ distance

6 ft. 8 in.

2,000 lb.

6 ft. 8 in.

2,000 lb. on inside footing

6 ft. 8 in.

1,000 lb. on corner footing

*A larger deck with multiple beams or cantilevers will affect the area supported.*

½ B to C | ½ B to C | ½ A to B | ½ A to B | Cantilever

C supports ½ distance to B.

B supports ½ distance to A plus ½ distance to C.

A supports cantilever area plus ½ distance to B.

C

B

A

Framing plans (see pp. 80-87) will specify the number of beams that support the deck, as well as the number of posts that hold up those beams. These specifications often assume good soil conditions. In the initial stages of foundation design, you can assume that the number and location of individual foundation points will be the same as the number of posts called for. Poor soil conditions, however, may require additional footings and posts.

Soil-bearing capacities can range from 400 psf (very poor) to 8,000 psf (very good). In soils with bearing capacities of 1,500 psf or more, which is typical of soils composed mainly of inorganic clays and silts or those that are sandy or gravelly, average footprint sizes should be fine. As the bearing capacity of the soil falls below 1,500 psf, as in soils with high amounts of organic clay, silt or peat, larger footprints will be needed to distribute the load properly. But increasing the footing size has practical limits, as large footings (those that are over 20 in. sq. for instance) need special engineering for strength and bigger holes in the ground to put them in. So, instead of larger footings, you can add extra footings to get the same results.

Keep in mind that some soils may be good enough when dry but will lose some of their bearing capacity when wet. For example, inorganic clays can expand when exposed to water and fine silts may get squishy. For this reason, it's important to keep footings and the surrounding soils dry by having proper drainage, which is discussed at right. Wet or low-bearing-capacity soil conditions can create complications that require the help of a structural engineer.

Your building department or state geological survey office should be able to help you determine what kind of soil you have. If all else fails and you're left with doubts, consult an engineer who will take core samples and run an analysis of the soil. If the deck is being built next to a house with a conventional and trouble-free foundation, chances are good that a conventionally sized deck foundation will more than fit the bill.

The deck shown in the drawing on p. 61 is moderately sized, with the center of the beam (and therefore the posts and footings) set back from the edge of the deck by 12 in. Note also that the centers of the end posts are set in 6 in. from each end of the beam, which is something I usually try to do for aesthetics. These cantilevers will change the area supported by each footing slightly and therefore the footing size compared to the simpler example in the drawing on p. 56.

In this example, let's assume that the bearing capacity of the soil is 1,500 psf (a pretty conservative figure). Our 12-ft. by 20-ft. deck (240 sq. ft.) is designed for the standard 50-psf load. Each middle post supports an area of 41.1 sq. ft., measuring 6 ft. 4 in. in a direction parallel to the beam (halfway to each neighboring post) and 6 ft. 6 in. perpendicular to the beam (1 ft. for the cantilever and 5 ft. 6 in. toward the ledger). Multiplying the area by 50 psf gives a 2,055-lb. load. In soil that can bear 1,500 psf, this calls for a 14-in. by 14-in. footing. Corner footings each support 23.8 sq. ft. (3 ft. 8 in. by 6 ft. 6 in.) and require smaller 11-in. by 11-in. footings. Footing sizes are rounded up to the nearest whole inch.

Once the size of the footprint has been determined, you need to choose the thickness of the footing. Six inches is adequate for most decks, but with heavier loads or footings over 16 in., footings should be 8 in. thick. The footings will not require reinforcing rebar as long as the distance that the footing extends laterally past the outside edge of the pier or post is not greater than the thickness of the footing. If rebar is needed, a grid with #4 bars laid 12 in. apart in both horizontal dimensions should be enough (see the drawing on p. 59), although you should verify this with an engineer.

The foundation design needs to address the potential for frost heave, which varies from region to region. Frost heave results from freezing temperatures, water and the ability of the soil to retain water. When water in the soil freezes, the soil expands with a force more than strong enough to lift the foundation of a deck. To avoid freezing conditions, the bottom of the footing must always be placed below the frost line, which is the maximum depth at which the ground will freeze in winter. This depth can be a few inches in moderate climates to as much as 48 in. where I build in Alaska. Your local building inspector will be able to tell you what is required in your area.

You can also minimize frost problems by allowing for good drainage below the footings. Keep downspouts from emptying near footings and make sure the ground is sloped away from them. Fine silt or clay soil needs to be kept particularly dry. If you know that the soil is likely to be wet much of the time because of a problem such as a high water table, it might be necessary to run a perforated drainage pipe surrounded by gravel along the bottom of the footings that empties into a suitable runoff point. Whenever you suspect that permanent or seasonal water will get to the footing it's a good idea to improve drainage by removing 12 in. of soil underneath the footing and replacing it with a compacted mixture of sand and gravel.

Another problem with footing-and-pier foundations in some regions is frost jacking, which happens when the upper part of a buried pier is grabbed by a ring of frozen soil and forced upward. The shift may be only a small amount each year, but it adds up. Breaking the bond between the column and the frozen soil will minimize the problem, and this can be done by wrapping the pier tube before burial with several layers of polyethylene sheeting. If the footing is cast integrally with the pier and is somewhat larger, as is usually the case, it acts as an anchor against the uplifting forces of frost jacking.

## From Footings to Above Ground

After sizing the footings, you need to bring the foundation above grade. There are several ways to do this. You can pour concrete piers (which is what I almost always do), you can use continuous posts or you can use precast concrete pier blocks.

**Concrete piers** Concrete piers are decay proof, strong and required by codes in many areas. In most cases, they can be cast integrally with the footings, and this is my preferred method. They can also be formed and poured later if the footings are large or complicated, but this requires two concrete deliveries instead of one. Piers should stick up several inches above grade to help keep water away from the ends of the posts they support, even if you're using rot-resistant posts. Codes require piers to be a minimum of 6 in. above grade if the posts they support aren't rot resistant.

Concrete piers that support 4x4 posts are usually 8 in. in diameter or 8 in. square. Larger posts require larger piers (10 in. to 12 in. diameter or square for 6x6 posts). Piers can be formed by using tubular fiberboard forms, such as Sonotubes, or by forms built with lumber. Piers are usually required by code to have steel reinforcement. A typical 8-in. below-grade pier might require one or two #4, L-shaped pieces of rebar that run vertically from the footing up through the pier (see the drawing on the facing page). If the pier is less than 8 in., a straight piece of rebar can usually be used.

When building on particularly good soils or rock, the footing may not need to be any larger than the pier. In that case, the pier form can simply be run to the bottom of the hole. But you still need to protect against frost heaves and jacking.

Sometimes it may make sense to run concrete piers all the way up to the beam. This eliminates short posts on low decks, but it requires that the metal bases that anchor the post to the pier be set at precisely the right height to anchor the pier to the beam. There will be little room for adjustments.

**Continuous posts** Instead of using a concrete pier, the connection from the footing to above grade can be made with a pressure-treated wood post. This post would extend from the footing to the beam, eliminating the joint at grade level. A continuous post can make for a more rigid structure and may be required in earthquake country or on decks that are more than 8 ft. tall. Additional stability can be provided by surrounding the post with concrete up to grade level, but be sure to slope the concrete to encourage drainage.

When using this method, be sure to place the uncut, factory-treated end of the post on the footing and keep the end grain protected from dampness by placing it on an asphalt shingle or sealing it with roofing tar. Placing 6 in. of compacted sand and gravel under the footing will also help keep the wood dry. I don't often use the continuous-post system for several reasons. For one thing, most clients are of the impression that wood below grade won't last as well as concrete. (Whether that's true or not with today's pressure-treated lumber is another matter.) Also, the deck is harder to repair if the post is damaged. Finally, it's difficult to align the posts accurately on a complex deck.

**Precast concrete pier blocks** Buried footings and the concrete piers that extend them above grade level are big and bulky and so are usually formed and poured in place. However, in warmer climates, where the frost-line isn't very deep, inexpensive precast pier blocks may be all that you need for a foundation. These 12-in. to 18-in. tall by 12-in. to 24-in. square blocks often come complete with a cast-in-place metal post base as well. It's a good idea to bury them as deep as possible, leaving the post base and several inches of concrete exposed, and to place them on 6 in. of compacted sand and gravel as well. It's easy to build forms and cast your own pier blocks if they're not available locally. A cheap way to make some nice round pier blocks is to buy a section of 12-in. diameter fiberboard concrete form and cut it into 12-in. sections. Stand these on a level surface, pour them full of concrete and add the post base.

Although precast blocks come with a variety of good post bases, one to avoid is the type with a block of wood cast into the top. This block, which often is not even pressure treated, is supposed to allow the post to be toenailed on, but it isn't adequate for even small lateral loads, and the blocks are prone to splitting.

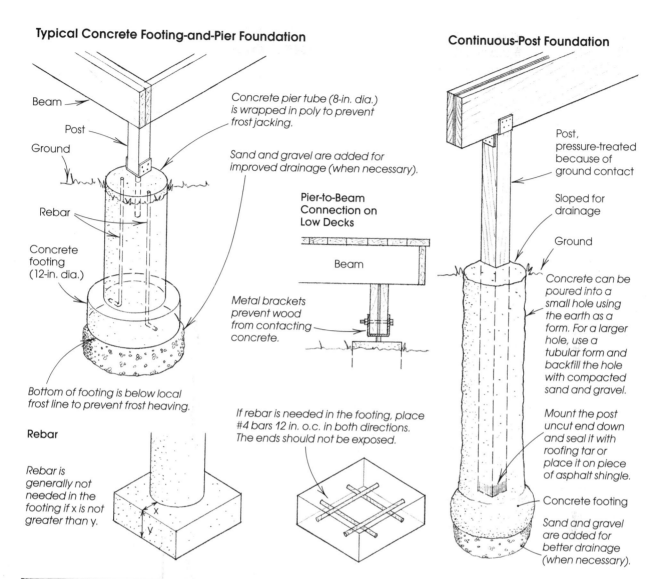

## Typical Concrete Footing-and-Pier Foundation

Beam

Post

Ground

Rebar

Concrete footing (12-in. dia.)

Concrete pier tube (8-in. dia.) is wrapped in poly to prevent frost jacking.

Sand and gravel are added for improved drainage (when necessary).

Bottom of footing is below local frost line to prevent frost heaving.

### Rebar

Rebar is generally not needed in the footing if x is not greater than y.

x

y

### Pier-to-Beam Connection on Low Decks

Beam

Metal brackets prevent wood from contacting concrete.

If rebar is needed in the footing, place #4 bars 12 in. o.c. in both directions. The ends should not be exposed.

## Continuous-Post Foundation

Post, pressure-treated because of ground contact

Sloped for drainage

Ground

Concrete can be poured into a small hole using the earth as a form. For a larger hole, use a tubular form and backfill the hole with compacted sand and gravel.

Mount the post uncut end down and seal it with roofing tar or place it on piece of asphalt shingle.

Concrete footing

Sand and gravel are added for better drainage (when necessary).

**Precast pier blocks with metal post bases already in place are an economical choice for shallow foundations. The one on the left provides a bigger footprint, which allows for more bearing surface. The block on the right can be buried a little deeper.**

## Layout

With all of the framing and foundation decisions worked out and clearly indicated on a set of plans, laying out the foundation becomes a simple matter of using some basic geometry to transfer the drawn plans to the deck site, this time using strings instead of pencils, and the earth as your paper. In all cases, begin by locating one side of the deck to act as the reference for the layout of the rest of the deck. Most typically this reference side will be the ledger board, which is attached to the house. I usually prefer to install the ledger before layout. If you don't want to install it now, you can reference your layout directly off the side of the house by adjusting your measurements to account for the ledger's future location. But if you locate and install the ledger as described on pp. 73-79, you'll have a fixed object from which to establish square corners and accurate measurements.

It's important to remember that all dimensions on the plans represent horizontal distances. So, whenever you use a tape measure at the building site, it must be kept as close to horizontal as possible. This can become particularly tricky on a sloped site where measurements can become severely distorted if you follow the slope of the ground rather than staying exactly horizontal.

Also, if the ledger is going high up on an exterior wall, such as on a second-story deck, you may want to find a way to avoid having to climb up and down a ladder to take measurements. I prefer to establish a reference string line using highly visible yellow mason's line about a foot off the ground, which lies directly under the face of the ledger. The string can be aligned using a plumb bob hung from the ledger and attached to nails or screws driven into the house siding. The ledger ends can be marked on the string with a felt-tip marker (see the drawing on the facing page).

The rest of the deck is located with strings, which will be attached to batter boards. A batter board is a 3-ft. or 4-ft. piece of 1x4 lumber attached about 1 ft. off the ground to a pair of 1x2 stakes driven into the ground (you can often buy pointed 1x2s at the lumberyard). String lines are then stretched tightly between batter boards to represent the location of the sides of the deck or the centerlines of the foundation footings. The strings are tied to screws or nails fastened to the top of the batter boards. Each corner of a deck without an existing solid reference like a ledger will need a pair of batter boards placed at right angles to each other; the corners are established where two strings cross. The batter boards should be located so that the strings will be approximately centered on

them, allowing plenty of room for adjustment. Keeping the batter boards 18 in. to 24 in. away from the intersecting strings will make them less likely to get bumped and knocked out of position when you are digging the holes.

Layout work is best done with two people because there's a lot of string stretching, measuring, adjusting and marking that has to go on simultaneously. As an example, we'll now lay out the same 12-ft. by 20-ft. deck that we used for sizing the footings.

The first string locates the common centerline of the footings and runs parallel with the ledger. First install a set of batter boards perpendicular to the house wall, and then stretch a string 11 ft. from the back of the ledger (or 10 ft. 10½ in. from the front) between them. Measure from one end of the string to the ledger and then the other, adjusting the string as necessary. Once it's the proper distance from the ledger on both ends, secure it to the batter boards.

Next, you want to establish each side of the deck by stretching strings from each end of the ledger to two more batter boards placed parallel with the ledger. If the deck is to be square, it is critical that these strings run square to the ledger and exactly parallel to each other. Carpenters traditionally check for square corners by applying the Pythagorean theorem, better known on the job site as the 3-4-5 rule: A triangle with sides measuring 3 ft. by 4 ft. by 5 ft. will contain a perfect right angle. Another way to check for square is shown in the drawing on the facing page. A rectangle is square if its facing sides are the same length and its diagonals are equal. If any of these dimensions is off by more than ⅛ in. for each 8 ft. of length, then the layout isn't as square as it should be.

I like to establish square corners by using a big homemade triangle, which my editor has generously dubbed "The Schuttner Square." I usually make one of these at the start of a job, and I find many uses for it through the course of a deck project. I make it out of straight 1x4s, as shown in the top drawing on p. 62. Once I've made the triangle, I just rest one side against the ledger (or any other reference surface) and run the string out parallel with the other side, making sure that no dips or bumps in the ledger throw off the alignment.

With the sides of the deck and the footing centerline laid out, you can now mark the exact locations for the outermost footings on the ground. In our example, measure 6 in. in from the side strings on the footing centerline string, then transfer this spot to the ground using a plumb bob and mark the spot on the ground with a nail and a piece of colored flagging or tape. Then locate the remaining footings along the

## Laying Out the Foundation

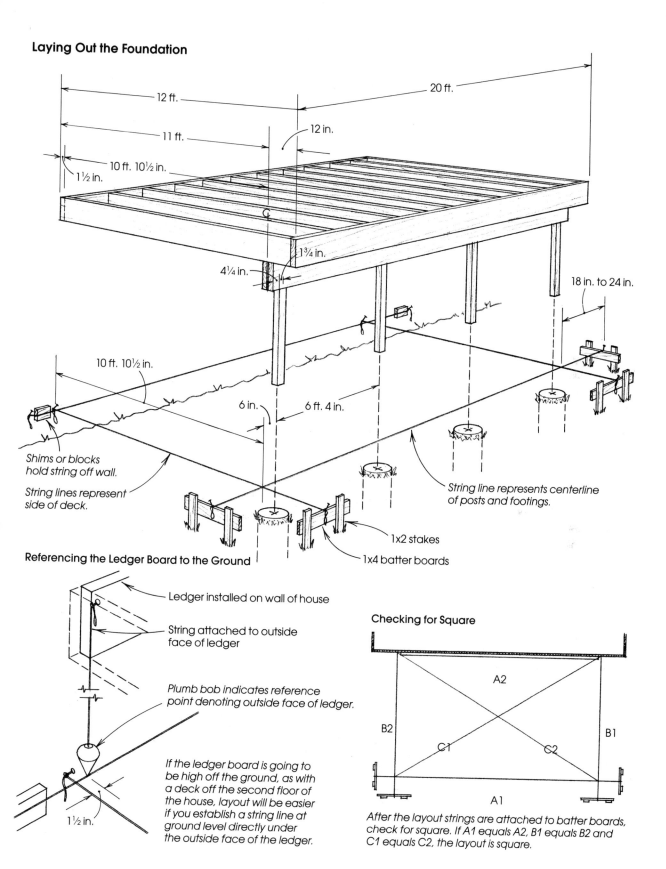

12 ft.

20 ft.

11 ft.

12 in.

10 ft. 10½ in.

1½ in.

1¾ in.

4¼ in.

18 in. to 24 in.

10 ft. 10½ in.

6 in.

6 ft. 4 in.

Shims or blocks hold string off wall.

String lines represent side of deck.

String line represents centerline of posts and footings.

1x2 stakes

1x4 batter boards

### Referencing the Ledger Board to the Ground

Ledger installed on wall of house

String attached to outside face of ledger

Plumb bob indicates reference point denoting outside face of ledger.

If the ledger board is going to be high off the ground, as with a deck off the second floor of the house, layout will be easier if you establish a string line at ground level directly under the outside face of the ledger.

1½ in.

### Checking for Square

A2

B2

B1

C1

C2

A1

After the layout strings are attached to batter boards, check for square. If A1 equals A2, B1 equals B2 and C1 equals C2, the layout is square.

## The Schuttner Square

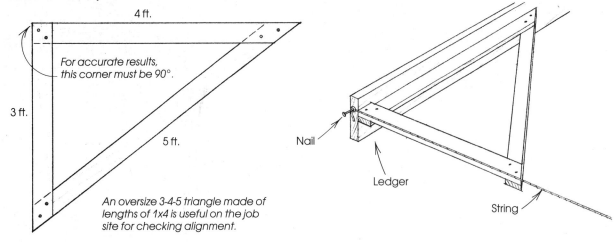

4 ft.

*For accurate results, this corner must be 90°.*

3 ft.

5 ft.

*An oversize 3-4-5 triangle made of lengths of 1x4 is useful on the job site for checking alignment.*

Nail

Ledger

String

## Laying Out Complex Decks

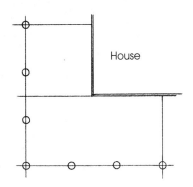

House

*On a deck that wraps around a corner of the house, extend a string line from one ledger out to the other string. By establishing two rectangles, you can easily check the layout for square.*

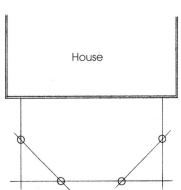

House

*For 45° corners, first run strings to establish a full rectangle, then measure back equal distances off of the outside corners to establish perfect 45° angles.*

*This deck can be laid out with a series of separate rectangles (A) or with one large rectangle with smaller rectangles referenced off of it (B).*

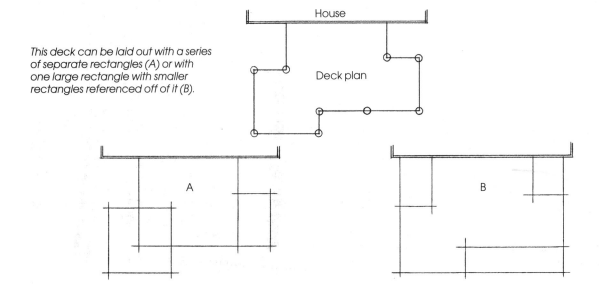

House

Deck plan

A

B

To determine the locations of the footings, measure along the footing-centerline string and use a plumb bob to transfer this position to the ground (top photo). Mark this spot with a nail and a piece of colored flagging or tape (above). Locate and mark all the footings in the same manner.

tape. Then locate the remaining footings along the string every 6 ft. 4 in.

If the deck has angles or offsets or is anything other than a simple rectangle, you'll need a lot more string lines. If there's more than one beam on the deck, there will be more than one line for footings. You may also need to choose reference surfaces other than ledgers, such as parts of the deck that are already represented by strings. If you don't have a ledger to reference from, as with a freestanding deck, you'll need to choose a location for a first string to act as the reference for the other sides. If the deck is made up of several separate structures, you'll first need to establish the four sides of one area and then use one of those sides as a reference for the other areas. Laying out for a 45° turn may involve establishing a square corner and then cutting across it with a diagonal string line. A deck plan with a lot of jogs might be laid out as a composite of separate rectangles, each referenced off another. Or it might be laid out as one large rectangle containing a series of subdivisions. Once the footings have all been located, the layout strings can be removed, but leave the batter boards in place because you'll need to recheck your work later.

## Digging Holes

If you've ever manually dug a hole 48 in. deep, you know that it can be hard work. Run into some large stones or bedrock and it can become a nightmare.

If you've got only a few holes to dig or if you're trying to cut costs wherever you can, the trusty old post-hole digger will probably get the job done. But you might want to investigate the price of drilling the holes before you start. I've had great luck hiring the owner of a local chain-link-fence company. He has a Jeep rigged up to drill holes for fence posts, and he moonlights evenings and weekends drilling holes for deck builders like myself. His little drill rig can drill up to 24-in. diameter holes 5 ft. deep if the ground isn't too rocky. And since it's on a four-wheel drive vehicle, it can go just about anywhere you'd want a hole. It cost me about $120 to have him drill the 25 4-ft. deep holes for the Morrises' deck , and it took a total of 1½ hours (including his travel) drilling in moderately rocky soil. I consider that to be money well spent. Track-mounted rigs could tackle even steeper or rougher grades, but they would cost a lot more.

Another option is to rent a hand-held gasoline-powered auger. These rigs are available in one-person and two-person models. They can generally dig holes only up to about 12 in. in diameter, maybe a little wider in soft soil. It's hazardous and brutal work, especially if the soil is hard and rocky, which causes a lot of abrupt stops and starts. But it's still quicker than the shovel.

Even if you use some power to dig the holes, you'll still probably want a manual post-hole digger to clean the loose dirt out of the holes. Remove as much loose soil as possible and then tamp the bottom of the hole hard with a post to insure good compaction. Remember that in soils with poor drainage you should dig the holes 6 in. deeper and add a layer of compacted gravel.

## A Footing-and-Pier Foundation

After digging the holes, it's time to construct a form system to contain and shape the concrete for the foundation. Using the sides of the hole as the form for the footing is expedient and works in most soil types except those that are loose or sandy, where it may not be allowed by local codes. If the hole is significantly larger than required for the footing, a square form can be constructed out of a 2x6s or 2x8s. This form should be held level and in the proper location by attaching it to 1x2 stakes driven around the form's perimeter.

Digging post holes can be the most grueling part of building a deck. On most projects, it's more cost effective to let a power rig do the digging.

Even if you dig the holes mechanically, you'll still need a manual post-hole digger to clean the loose dirt out of the bottom of the hole.

Because the hole is usually wider than the pier diameter, the pier requires its own form. I like to pour footings and piers in one step, which is easy to do using tubular fiberboard forms—Sonotube is probably the best-known brand. These forms are widely available, lightweight, cheap, leakproof and easy to cut to length with a handsaw. They come in 12-ft. lengths and cost me about a dollar a foot in the 8-in. diameter size. They're waxed on the outside, but don't leave them lying around the job site in the rain. Once the concrete is poured and cured, the exposed portion of the tube can be removed down to ground level.

The tubes should extend from the top of the footings to several inches above grade (your code may stipulate just how many inches; I usually shoot for 2 in.). It's not important that the tops of all the piers end up being level with each other. The wood posts can be cut to different lengths to compensate for differences in distance between pier brackets and the bottom of the beam. (They should also be cut as square as possible.) If you're building on particularly good soil, your footings may not have to be any wider than the piers. But in most cases, you'll need to suspend the tubes off of the bottom of the hole so that concrete can spread out and form a larger footing (see the drawing on p. 68). The bottom of the tube would then be level with the top of the footing. The tube can be held up by attaching it with screws or nails to a pair of staked 2x4s that span the hole. When the concrete is poured into the tube it can be encouraged to spread out to the sides of the footing form by tamping it with a 2x4. The concrete will spread more easily if it's not too stiff. This

# Installing a Footing-and-Pier Foundation

1. Pier tubes can be cut easily with a handsaw, but they should be cut square. Cut them long enough to reach from the top of the footing to about 2 in. above grade.

2. To help minimize frost jacking, wrap the tubes with poly before dropping them in the holes. In all but the best of soils, you'll need to suspend the tubes off of the bottom of the hole so the concrete flows out the the bottom to form wider footings.

method works fine if the footing is not more than twice the pier diameter. If the hole in the ground is large enough, concrete can be poured directly into the form. With larger footings and large holes, you'll want to stabilize the bottom of the pier tube in the center of the footing, as shown in the drawing on p. 68.

## Handling Concrete

My first construction job was feeding a concrete mixer for several months at a remote site where a contractor was building a swimming pool. I don't remember it as a miserable experience, but they say we have no memory for pain. After all these years, I still experience a certain hypnotic fascination watching the blades of a mixer beat the ingredients into a grey soup.

Concrete is a mixture of aggregate (sand and gravel), water and cement. Most concrete mixing directions state that the water should be clean enough to drink, and while this need not be taken literally, the water and the aggregate do need to be free of dirt, debris, salt and other contaminants. The sand should be a mix of fine particles smaller than ¼ in. in diameter. The gravel should be a mix of sizes from ¼ in. to 1½ in. My local gravel yard has a pile of premixed aggregate in about the right proportions, but the ingredients are more commonly sold separately. You can purchase sand and gravel along with bags of cement and mix them together yourself, or you can buy bags of premixed concrete to which you add water. But the

4. The footings here are the same diameter as the piers, so the holes are backfilled with the dirt that came out of them. (If there's a ground-water problem, the hole should be backfilled with a mix of compacted sand and gravel.) Put in about 8 in. of dirt, then tamp it down. Repeat this process all the way to the top. If you used wood forms instead of tubes or have a footing larger than the pier diameter, let the concrete harden overnight before removing the forms and backfilling the hole. Be careful not to shift the top of the pier. As you backfill the holes, make sure to keep the tube level.

3. The layout strings will help you to center the tubes in their holes. You may need to remove the strings when it comes time to pour concrete, but you'll want them back in place to set the post bases.

premixed bags yield only about ⅔ cu. ft. and cost enough to make them practical only for small jobs.

Concrete hardens in a chemical process between the water and cement called hydration. Hydration requires only a small amount of water, but the mix usually needs more to allow the concrete to slide down a chute or pour from a bucket. However, too much water can weaken the concrete. The consistency of freshly mixed concrete is measured with a "slump test." The concrete is poured into a truncated cone, and then the cone is removed, allowing the concrete to settle (or "slump"). If it settles 3 in., then the mixture has a slump of 3 in. (concrete is often ordered with a slump of 4 in.) The rule of thumb is that concrete should be as dry and stiff as possible while still being able to fill all the voids in the form. Soupy is too wet.

The strength of the concrete (its "compressive strength"), which is measured in pounds per square inch (psi), is determined by the amount of cement in the mix. The standard mix uses five bags of cement per cubic yard and yields a compressive strength of about 2,500 psi, which is adequate for most deck foundations. A mix of 5½ bags per cubic yard yields a strength of 3,000 psi, which may be necessary to handle a heavier load.

One decision you need to make is whether or not to mix your own concrete. Depending on your pace, you can mix between one-half and one cubic yard of concrete a day in a wheelbarrow, one batch at a time. But mixing concrete is hard work and if you need more than one cubic yard, you might want to cut your effort and time in half by renting a mixer. Bear in

5. Drop a length or two of rebar (depending on your local code) in each hole. After you've poured about a foot of concrete into the hole and the concrete has had a few minutes to harden, pull the rebar up so that it's about 4 in. off the bottom and 2 in. below the top of the form.

6. Once the tube is filled with concrete, smooth off the top (a flexible pancake turner works well) and drop in a post base, wiggling it around to ensure that it is surrounded by concrete.

## Suspending and Stabilizing Pier Tubes

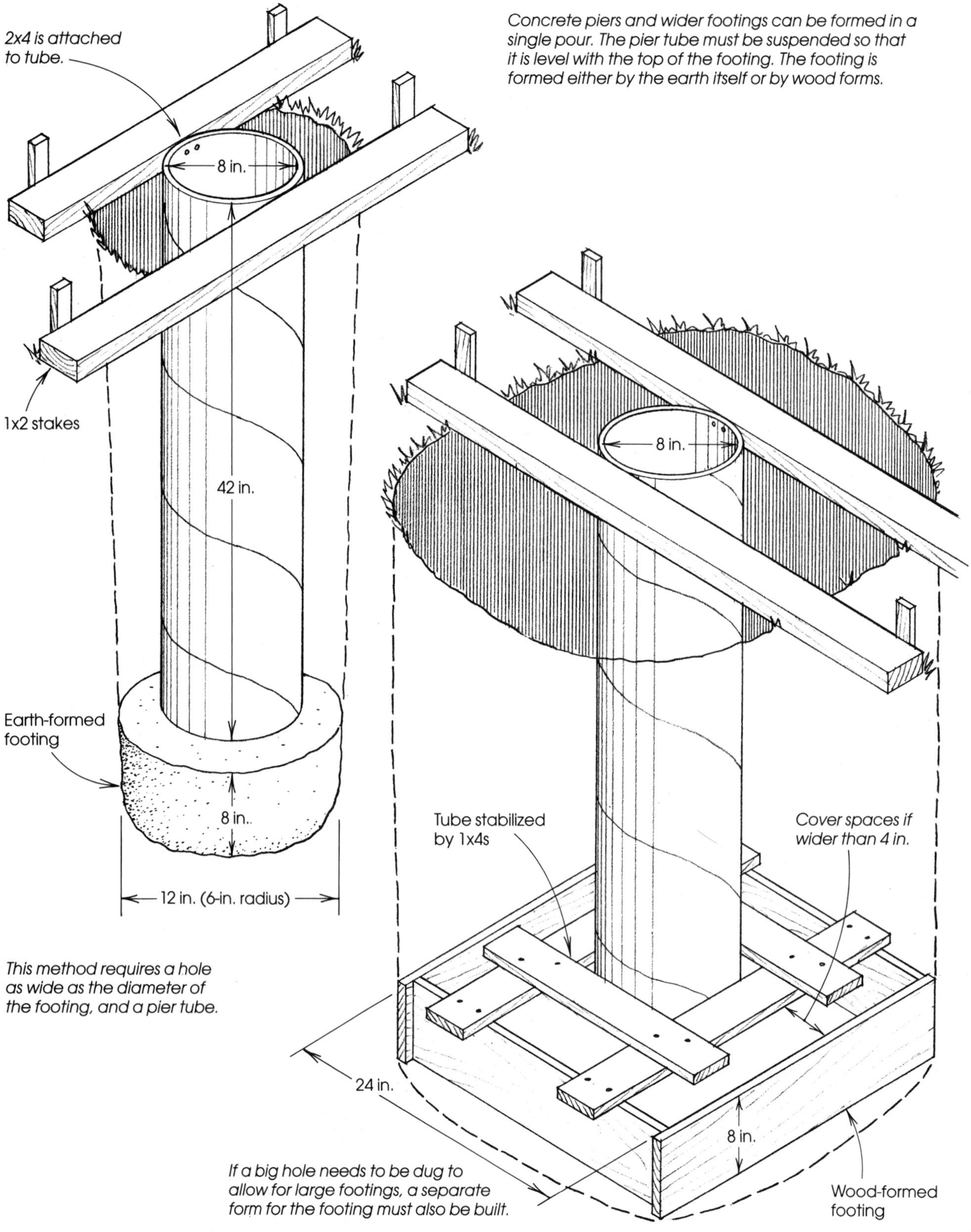

*2x4 is attached to tube.*

*Concrete piers and wider footings can be formed in a single pour. The pier tube must be suspended so that it is level with the top of the footing. The footing is formed either by the earth itself or by wood forms.*

8 in.

1x2 stakes

42 in.

8 in.

Earth-formed footing

12 in. (6-in. radius)

8 in.

Tube stabilized by 1x4s

*Cover spaces if wider than 4 in.*

*This method requires a hole as wide as the diameter of the footing, and a pier tube.*

24 in.

8 in.

*If a big hole needs to be dug to allow for large footings, a separate form for the footing must also be built.*

Wood-formed footing

# Estimating concrete

When you order a delivery of ready-mix concrete, you'll need to specify the quantity, the compressive strength and perhaps the slump. Concrete is sold by the cubic yard, which is 3 ft. x 3 ft. x 3 ft., or 27 cubic feet. Often you'll have to pay for a 1-cu.-yd. minimum delivery even if you need less. Estimating the quantity of concrete needed is a simple matter of figuring out the amount needed for each footing and pier and then multiplying by the total number required. Using tubes on deck jobs makes estimating concrete needs much easier. Here are some useful formulas for estimating volume:

1,728 cu. in. = 1 cu. ft.
27 cu. ft. = 1 cu. yd.
radius = ½ diameter
**For cylinders:** volume = $\pi r^2 h$, where $\pi = 3.14$, r = radius of the cylinder and h = height.
**For boxes:** volume = h x w x l

As an example, to figure the volume of concrete needed for the foundation shown in the drawing at left on the facing page, first calculate the volume of the 8-in. dia. x 42-in. long pier cylinder:

$\pi \times 4^2 \times 42 = 2{,}110$ cu. in. per pier.
2,110 cu. in. ÷ 1,728 cu. in. = 1.22 cu. ft. per pier.

Next, calculate the volume of the 12-in. dia. x 8-in. thick footing:

$\pi \times 6^2 \times 8 = 904$ cu. in. per footing.
904 cu. in. ÷ 1,728 cu in. = 0.52 cu. ft. per footing.

Now multiply each volume figure by the number of piers and footings and add them together to get the total volume of concrete needed. (Add a little extra to account for spillage and error.) For four piers and footings:

1.22 x 4 = 4.88 cu. ft. for piers
.52 x 4 = 2.08 cu. ft. for footings

4.88 cu. ft. + 2.08 cu. ft. = 6.96 cu. ft. total volume

For the square footing shown at right on the facing page, calculate volume as follows and multiply by the number of piers and footings needed:

24 in. x 24 in. x 8 in. = 4,608 cu. in. per footing.
4,608 cu. in. ÷ 1,728 cu. in. = 2.67 cu. ft. per footing.

mind that the raw materials for half a cubic yard of concrete may strain the carrying ability of a normal pickup truck.

If you need more than a yard and your time means anything to you, then consider calling the ready-mix people. The cost of labor for mixing and gathering up all the ingredients and tools means that on an average deck job it isn't cost effective to have my crew make concrete. Even if the concrete company adds a surcharge for a small delivery, I still come out ahead. But you'll lose money fast if you aren't ready to pour the concrete when the truck arrives. All the holes should be dug, the forms in place, wheelbarrows and shovels at the ready and the crew finished with their coffee break.

Hand mixing concrete is straightforward work whether it's done with a mixer or with a hoe in a wheelbarrow. The standard mix is three shovels of gravel, two shovels of sand and one shovel of cement. Try and keep the shovelfuls of equal size and remember that it's better to have too much cement rather than too little. Throw this amount into a wheelbarrow and then mix the dry ingredients well with a hoe—a real concrete hoe has a hole in the middle of the blade, but the garden variety will work fine. Shovels don't work well for mixing, and they're a lot more work.

After the dry ingredients are mixed, start adding water—just a little at a time. Keep mixing and adding water until the mix is just wet enough to fill the voids in the forms easily, but don't let it get runny. If it's too soupy, just add more dry goods. Wear gloves, and rubber boots are essential if you're the messy type or plan on wading through the concrete. When you're done, wash off all the tools as soon as possible.

Once the concrete has been poured, it needs to cure slowly, and that requires that it stay wet. Concrete that has been poured below grade will usually be all right, and the tops of piers need to be covered only during very hot and dry conditions. Let it rest overnight before working on it.

# Framing

## Chapter 4

The framing is the skeleton of the deck, on which we hang the decking, railings and accessories. For the most part it's hidden and unglamorous. Because the framing needs to provide support for many years, and because it can be difficult to repair after it's been built, it needs to be constructed solidly out of good materials and installed with close attention to detail.

The principal components of a deck frame are joists, ledgers, beams and posts. Framing lumber is most often pressure-treated Douglas-fir, Southern yellow pine or hemlock, although weaker woods like redwood are occasionally used. If you have more than one species of framing lumber available, you'll need to weigh any price differences against any changes in size required by differences in structural strength.

Joists are the horizontal members that lie directly under the deck boards. Joists are usually attached on one end to a ledger on the house, and they stretch out to a rim joist that caps the other end. The rim end of the joists is supported by a beam running perpendicular to the joists. The beam rests on posts that in turn are set on the foundation.

Joists and ledgers are usually made of 2x lumber. Joist size is determined by the load the joists must carry and the length of their unsupported span. If a large span requires a joist size that is unavailable or too expensive, you can add additional support in the form of an extra beam to break the span in half. That allows you to use narrower joists, but it also requires another set of posts and footings.

Beams are made of either solid timbers or built-up 2x framing lumber. Beam size is determined by the deck's size, how many beams are needed to support the joists, and the beam's span between posts. Adding more posts will decrease the beam size but increase the foundation work. Beams can be located in the same plane as, or underneath, the joists.

Finally, the size of the posts is determined by how many are used, that is, by the load that each must bear and their length. As explained in the previous chapter, the number of posts needed can also be affected by the bearing capacity of the soil.

Choosing the correct size of joists, beams and posts on a small deck is straightforward, but each framing decision affects the next—the number of the posts will affect the size and span of the beam and so forth. Your first decisions may be tentative and change as you refine the overall framing plan. This chapter will help you find the right balance of economics, aesthetics and good engineering.

Some decks are not so straightforward. Level changes, angles, curves and incorporating special objects such as trees or hot tubs are some of the features that make a deck unique, but they require special framing, planning, labor and material. Occasionally someone even wants to put a deck on a roof, and I'll give some help on that, too.

This chapter will also help you choose the right size and type of materials and explain the correct way to install them. All components must be placed to shed water away from the house proper, and the connections between the components must be reliable and long lived.

## Two Ways to Frame a Deck

### Beam-Over-Post Framing

*The cantilever should not exceed one-fifth of the maximum rated span of the joists. It should never exceed one-half the length of the joist that is supported between the ledger and the beam.*

Joists spaced 16 in. or 24 in. on center, typ.

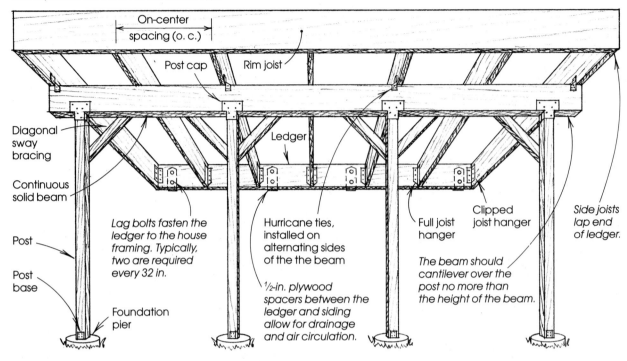

On-center spacing (o. c.)

Post cap

Rim joist

Diagonal sway bracing

Continuous solid beam

Ledger

Post

Post base

Foundation pier

*Lag bolts fasten the ledger to the house framing. Typically, two are required every 32 in.*

*Hurricane ties, installed on alternating sides of the the beam*

*½-in. plywood spacers between the ledger and siding allow for drainage and air circulation.*

Full joist hanger

Clipped joist hanger

*The beam should cantilever over the post no more than the height of the beam.*

*Side joists lap end of ledger.*

### Continuous-Post Framing

*Continuous posts form railing.*

*Joist should slope ¹⁄₁₆ in. per foot away from the house.*

*When the ledger can be firmly attached to the house only by cutting the siding, flashing must be added to keep out water.*

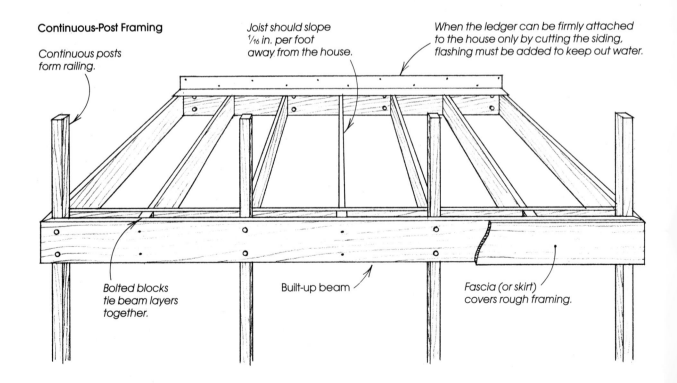

*Bolted blocks tie beam layers together.*

Built-up beam

*Fascia (or skirt) covers rough framing.*

The ledger is attached far enough under the threshold to allow for decking to be installed, plus an extra ¾ in. to keep water away from the threshold.

The ledger should be installed at different levels to allow for a level change in the deck. In this case, an intermediate step bridges the two levels.

The ledger must be continuous around a corner. Here, the double joists allow for the deck boards to be mitered at the corner.

## The Ledger

The ledger is a horizontal board that is attached directly to the house and supports one end of the joists. It is the first component to be installed and acts as a reference for the other parts. I find it helpful to install the ledger before doing the layout and foundation work. The ledger is really nothing more than a rim joist (the joist perpendicular to and capping the ends of all the regular joists) fastened to a wall of the house. Because it carries a large part of the deck load, it needs to be bolted solidly to the framing or foundation of the house. Depending upon the exterior surface material of a house, the ledger can be put on over the siding or set into an opening cut into the siding. Both methods use special flashing techniques to guard against potential moisture problems.

**Sizing the ledger**   The ledger is the same size and made of the same lumber as the joists, which it usually supports with metal hangers. If all the main joists are 2x10 pressure-treated Southern yellow pine, then so is the ledger. Pick out some of the straightest joists from the pile to use for the ledger stock—the rest of the building goes more easily if you can start off with a flat face and a straight top edge. Try to keep the joints in the ledger to a minimum by using long pieces: two 12-footers would be better than three 8-footers.

Ledgers can also support beams instead of joists if the decking pattern requires the beams to run into the ledger (see "Special Framing Techniques" on p. 94). In this case, the ledger may need to be thicker than a 2x to accommodate metal beam hangers.

**Locating the ledger**   The ledger is located in the same plane as the joists, and this height should have been determined in the design stage. The top of the ledger must be lower than the bottom of all doorway thresholds by at least the thickness of the decking. It's a good idea to drop the ledger an additional ¾ in., which creates a small step up to the bottom of the threshold. So, if you're using 2x decking, the top of the ledger should be 2¼ in. below the bottom of the threshold (¾ in. plus 1½ in). If you're using 5/4 decking, the distance would be 1¾ in. When flashed as described later, this drop will ensure that rain won't be wicked in under the threshold.

Any drop greater than ¾ in. will create a potential stumbling point and is unnecessary for deterring water. If you're concerned about snow blocking an outward swinging door, you can make the drop larger. A better solution, however, and one that must be made at the design stage, is to drop the entire level of the deck by a whole step of about 6 in. and add a narrow landing at least 11 in. wide at threshold level.

Normally, the ledger is as long as the deck, with one adjustment. The end joists should lap over the end of the ledger rather than butt into it like the others. This allows the end joists to be nailed to the ledger (a regular joist hanger won't work here), and it looks better to cover the end grain of the ledger. Therefore, the ledger should be shortened 1½ in. at each end to provide room for the end joists (as shown on p. 72).

To lay out the location of the ledger, measure down from the bottom of the threshold the thickness of the decking plus ¾ in. and make a mark. Next, extend this mark horizontally to represent the top of the ledger. To make the layout line perfectly level, use a 4-ft. level attached to a long, straight board or a water level. Mark spots A and B, as shown in the top drawing on the facing page and then connect the marks with a chalkline. Next, mark vertical lines indicating the ends of the ledger with a square, remembering that the ledger is one joist-width shorter on each end than the width of the deck.

**Attaching the ledger** The ledger carries a large part of the load of the deck, so it must be firmly attached to the house with bolts that won't loosen over time. Most often the ledger will be bolted to the framing of the house, but on some decks it may need to be attached to a concrete foundation or a brick wall. Since the deck is usually built at about the same level as the house's floor, the ledger is usually attached by bolting into the solid rim or band joist of the house's floor-joist system. Attaching to studs in a wall is also acceptable if they're the only framing available at the deck's level. Bolting into concrete is usually fine if the deck level falls into the foundation region. It would not be acceptable to bolt only into siding or thin plywood or board sheathing.

There are two principal methods of attaching the ledger to the house—the one you use will depend on the type of siding on the house. For the most part, you'll use the first method on a house with a relatively flat exterior and the second method on a house with an irregular surface.

The first method requires no tampering with the siding. The ledger is held away from the siding by plywood spacers, which create a gap through which water can drain. This method requires that the surface be strong enough to hold a bolted ledger, such as concrete or brick, or that the siding be flat against a substrate like the framing and sheathing without gaps that would cause the siding to crack or crush when the ledger bolts are tightened down. Tongue-and-groove, shiplap, plywood, stucco, some types of rabbeted horizontal bevel siding, and board-and-batten siding would be appropriate for this approach.

I make pointed spacers out of ½-in. pressure-treated plywood. They should be at least 3 in. wide and about 2 in. shorter than the width of the ledger. Installed point up, they allow for good drainage. The ledger will be bolted through the spacers, so you'll need spacers at each bolting location (usually 32 in. o. c.). It's a good idea to make gaskets out of 90-lb. roofing felt and place one on each side of the space. The gaskets will help prevent water from getting to the bolt and wicking into the house.

Some types of siding rest flat against the sheathing or house framing on the backside, but are not really smooth on the front (horizontal-bevel siding with rabbeted bottoms is the main example). This siding requires a specially shaped spacer (shown in the lower right drawing on the facing page), which allows for a solid fit between siding, spacer and ledger. Make these spacers out of pressure-treated wood and soak them in preservative to ensure that the cut edges are protected from the elements.

If you need to bolt to studs that don't coincide with the desired bolt pattern, you'll have to add extra spacers and rearrange bolt locations to allow for the required number of bolts. It never hurts to have a few extra, if you have any doubts. The spacers should be located about an inch below the top edge of the ledger. Tack the spacers in place with one or two 8d nails. Once they're installed, the bolts will hold them permanently. With the layout complete and the spacers in place, the ledger can be bolted to the house.

The second method for attaching the ledger to the house is more complicated, but it's necessary on houses with clapboard, wood shingle, unrabbeted horizontal bevel siding, aluminum, vinyl and similar types of siding. This siding does not rest flat against the sheath-

## Locating the Top of the Ledger on the House

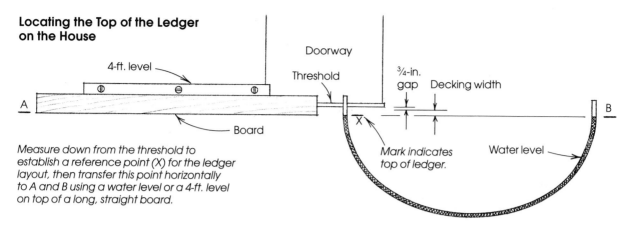

4-ft. level

Doorway

Threshold

¾-in. gap

Decking width

A

Board

X

Mark indicates top of ledger.

Water level

B

*Measure down from the threshold to establish a reference point (X) for the ledger layout, then transfer this point horizontally to A and B using a water level or a 4-ft. level on top of a long, straight board.*

## Attaching the Ledger to a Smooth Surface

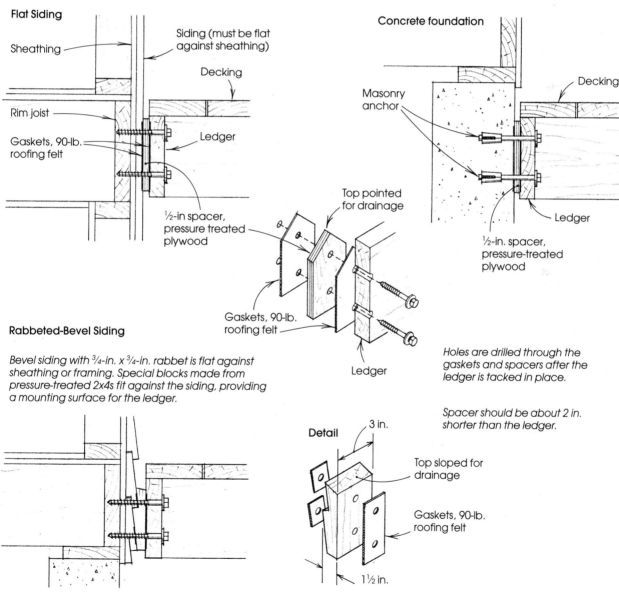

### Flat Siding

Sheathing

Siding (must be flat against sheathing)

Decking

Rim joist

Ledger

Gaskets, 90-lb. roofing felt

½-in spacer, pressure treated plywood

### Concrete foundation

Masonry anchor

Decking

Ledger

½-in. spacer, pressure-treated plywood

Top pointed for drainage

Gaskets, 90-lb. roofing felt

Ledger

*Holes are drilled through the gaskets and spacers after the ledger is tacked in place.*

### Rabbeted-Bevel Siding

*Bevel siding with ¾-in. x ¾-in. rabbet is flat against sheathing or framing. Special blocks made from pressure-treated 2x4s fit against the siding, providing a mounting surface for the ledger.*

*Spacer should be about 2 in. shorter than the ledger.*

Detail

3 in.

Top sloped for drainage

Gaskets, 90-lb. roofing felt

1½ in.

## Attaching the Ledger on an Irregular Surface (After the Siding Is Removed)

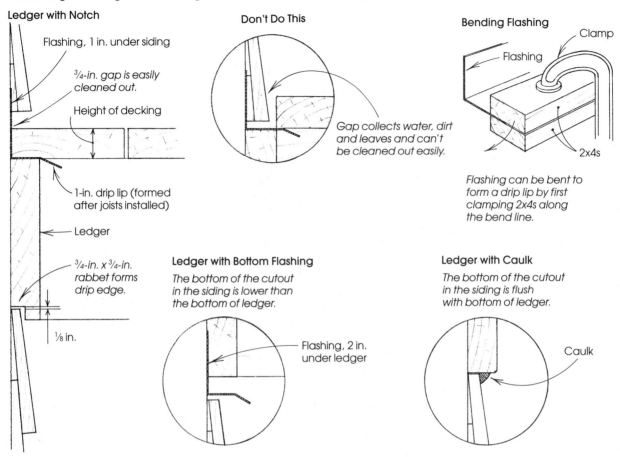

**Ledger with Notch**

Flashing, 1 in. under siding

*¾-in. gap is easily cleaned out.*

Height of decking

1-in. drip lip (formed after joists installed)

Ledger

¾-in. x ¾-in. rabbet forms drip edge.

⅛ in.

**Don't Do This**

*Gap collects water, dirt and leaves and can't be cleaned out easily.*

**Bending Flashing**

Clamp

Flashing

2x4s

*Flashing can be bent to form a drip lip by first clamping 2x4s along the bend line.*

**Ledger with Bottom Flashing**

*The bottom of the cutout in the siding is lower than the bottom of ledger.*

Flashing, 2 in. under ledger

**Ledger with Caulk**

*The bottom of the cutout in the siding is flush with bottom of ledger.*

Caulk

**Don't Do This!**

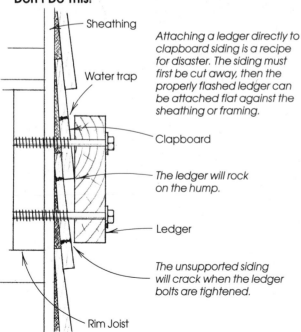

Sheathing

Water trap

Clapboard

*The ledger will rock on the hump.*

Ledger

*The unsupported siding will crack when the ledger bolts are tightened.*

Rim Joist

*Attaching a ledger directly to clapboard siding is a recipe for disaster. The siding must first be cut away, then the properly flashed ledger can be attached flat against the sheathing or framing.*

ing or house framing, which makes it difficult to tighten the ledger bolts without damaging the siding. The solution is to cut out a section of the siding, thereby exposing a flat backing for the ledger. Once you cut into the siding, however, you've created a potential entry for water that must be countered with careful attention to flashing.

Locate the top of the ledger as described on p. 74. Then measure back up about ½ in. to mark the cutout line. This gap will keep the siding from resting on the decking, as shown in the top left drawing on the facing page. Now extend this mark horizontally, using a level (shown in the top drawing on p. 75) and snap a chalkline to establish the top cutline.

The cutout at the bottom of the ledger can be handled in several different ways. The simplest, neatest method, which eliminates the need for metal flashing, involves cutting a ¾-in. by ¾-in. rabbet in the bottom back edge of the ledger (see the drawing above left). This creates a lip that overhangs the siding and helps keep water out. With this method, the siding should be cut out ⅝ in. higher than the bottom of the ledger. The joists that lap either end of the ledger will need to be notched on the bottom as well. Caulk the joint where these joists meet the ledger as well as the siding. Also caulk any butt joints in the ledger. This method will hide an uneven cut in the bottom siding.

If the back side of the ledger is recessed into the siding more than an inch or so, cutting a rabbet would leave only a small and delicate lip. Then metal flashing will be a more durable option. For this, cut out the siding ¾ in. below the bottom of the ledger (center drawing, facing page) and add flashing between them. The flashing fits behind the ledger, covers the siding and is bent over to create a drip edge.

If the ledger is in an area that won't receive much moisture, such as under a large roof overhang, the rabbet and bottom flashing are unnecessary (drawing at right, facing page). The opening in the siding can be cut even with the bottom edge of the ledger. The flashing on top of the ledger should keep most water from reaching the bottom. The joint between the bottom of the ledger and the siding should be carefully caulked.

After laying out the top and bottom cutlines on the siding, mark the vertical cutlines with a square. Keep in mind that the cutout must be wide enough for the ledger and the overlapping joists on each end.

The next step is to cut the siding. This can be done by first removing whole pieces of siding, cutting them and then replacing them. Removing old siding is tedious and risky, and is usually only worthwhile with wood shingles, which are full of nails and irregular bumps. To remove old siding, work from the bottom up and pry the siding just enough to back the nails out a little. Then push the siding back down and remove the nail. Carefully cut the siding along the layout lines and reinstall.

I prefer to cut the siding in place. Cutting straight lines to a consistent depth on irregularly surfaced siding can be a real challenge. You first need to create a flat surface upon which to slide the base of your saw. This is done by tacking a 1x4 to the siding, spanning several courses of the siding. Remove any nails in the path of the cut and set the depth of cut so that you don't cut into the sheathing or framing.

Start the cut by positioning the blade over the cutline and temporarily holding the blade guard up. Start the saw, drop the spinning blade onto the cutline and then release the blade guard. You should be well braced and use caution with a plunge cut such as this. Don't overcut the layout lines. Instead, use a chisel to finish the cuts.

When making vertical cuts on uneven siding, a 1x4 guide provides a flat surface for the saw to ride on.

To ensure neat corners, finish cutting the siding with a chisel.

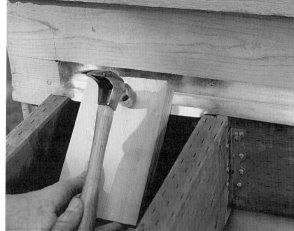

The side joists should lap over the ends of the ledger, with flashing installed to cover the joint between them (above). Once the joists have been installed, the flashing can be bent into a drip edge using a wood block and a hammer (above right). The flashing must be notched to fit tightly around the threshold (right). Caulk the joint between the threshold and the flashing after the joists have been installed.

Now that you've cut a hole in the siding, you need to turn your attention to flashing to keep water out of it. The flashing should be installed before the ledger is put on. I use L-shaped galvanized flashing, which I buy in 10-ft. lengths. On the top flashing, the vertical leg must be long enough to slide under the siding by an inch and reach the top of the ledger. The horizontal leg needs to cover the top of the ledger and extend an extra inch or so, which will be bent to form a drip edge later. The drip edge is essential to keep water off the face of the ledger, which in turn minimizes water reaching the bottom of the ledger.

The flashing goes from one end of the ledger opening to the other so that it extends over the end joists once they're in place. To install the flashing, put it in the hole with a leg under the siding and push the whole piece vertically up as far as it will go to get it out of the way before installing the ledger. Once the ledger has been installed, pull the flashing down into final position on top of the ledger. Overlap and caulk all butt joints in the flashing. After the joists have been installed, you can form the drip edge on the top by bending the flashing with a hammer and a block of wood, as shown in the top right photo above. If nails prevent you from pushing the flashing behind the siding and you can't easily remove them, you'll have to cut slots in the flashing.

If you're flashing the bottom of the ledger, the flashing should run up behind the ledger about 2 in., cover the ½-in. gap between the ledger and the siding, then overhang the siding by about an inch and end in a drip edge (center drawing, p. 76). You may get a better looking job if you form the drip edge on the bottom flashing before installing it. This can be done by clamping the drip-edge section between a couple of 2x4s and then bending it.

Before installing the decking, caulk the ends of the flashing where it butts into the siding. It's also a good idea to caulk the top of each joist under the flashing.

Thresholds need to be flashed with special care. If you're adding a new door out to the deck, you should cut the flashing to fit the doorway and then bend the flap to fit under the threshold before it's installed. Otherwise, cut the flashing so that it fits snugly around the threshold and then caulk the joint.

**Bolting the ledger**  The ledger is bolted to the house in much the same way whether you cut out the siding or are using spacers. Lag bolts are the most common fasteners used, but in the unlikely event that you have access to the nut side you'll get a stronger connection using standard hex bolts and nuts.

First, locate the solid house framing. If the deck is at floor level, chances are you'll be bolting to the rim joist. But if you need to bolt into wall studs, you'll have to locate them by probing with a nail and hammer. Nail heads on the siding are a good indication of solid framing underneath.

Take a few moments to check the joist layout on the ledger. You don't want to put bolts behind deck joists because that will make them unavailable for retightening in the future. The bolt may also interfere with the joist hanger. Plan on using two vertically aligned 7/16-in. bolts every 32 in., and space them every 24 in. if the joists span over 10 ft.

Lag bolts should be long enough to pass through the ledger and any spacers, siding and sheathing and penetrate the house framing at least 1½ in., with an extra ½ in. poking through the backside of the framing to maximize the grip. If possible, add extra blocking behind the rim joist and use even longer lag bolts (shown below).

The holes for lag bolts need to be predrilled, using spade bits or augers. Drill the larger shank hole for the unthreaded part of the bolt first, then extend it with a smaller pilot hole for the threads. The depth of the full-size shank hole should be as long as the length of the bolt without threads. To allow for a tight fit, the second pilot hole should be slightly smaller than the diameter of the shaft on the threaded section of the bolt and be as long as the bolt. Insert the lags with washers and tighten with a socket and ratchet. The bolt should be tight in the hole. If it continues spinning as you tighten it, then the pilot hole is too big. Caulk under the washer before you tighten it to provide additional insurance against any water penetrating along the bolt.

**Attaching to masonry**  If the ledger needs to be attached to concrete or brick, you need to choose an appropriate masonry fastener (see the discussion on p. 33). A range of fasteners can be used in solid concrete and with bricks that are in solid condition and firmly attached to the house. But if the bricks are old and soft, you can't rely on a masonry fastener to support the ledger. In this case, it may be possible to drill through the brick to some solid framing underneath, but this requires exceptionally long lag bolts. You may want to avoid the problem by building a freestanding deck that doesn't attach to the house at all, but this requires a full foundation system on both sides of the deck.

Bolts should be spaced in masonry the same as in wood. You can get by with ½-in. wedge or sleeve anchors with nuts or ½-in. lag bolts with masonry shields. Follow the manufacturer's advice on determining the correct length of fastener to use.

Drill the bolt holes in the ledger, then temporarily fasten or hold the ledger in place. Using the ledger

## Bolting the Ledger

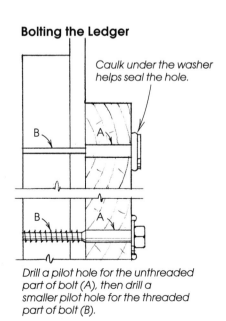

Caulk under the washer helps seal the hole.

Drill a pilot hole for the unthreaded part of bolt (A), then drill a smaller pilot hole for the threaded part of bolt (B).

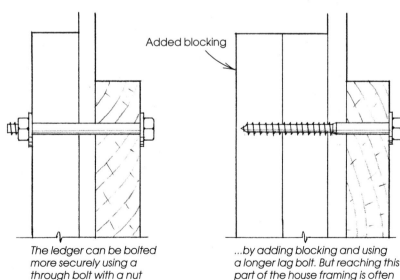

Added blocking

The ledger can be bolted more securely using a through bolt with a nut on the backside or...

...by adding blocking and using a longer lag bolt. But reaching this part of the house framing is often difficult or impossible.

holes as guides, drill the holes in the masonry with a masonry bit. The holes in the ledger should be the same diameter as the fastener. You can drill the holes with a masonry bit and a regular drill if you've got only a few to do—with any more than that you should rent a hammer drill, which will work much faster. Put the fastener inserts into the holes, and then bolt the ledger and spacers to the wall. If you find that some bolt holes in the ledger aren't aligned properly, you'll have to ream the holes somewhat oversize.

## Posts and Beams

Posts and beams are the backbone of the deck's structural skeleton. Joists are supported by perpendicular beams, and the beams are supported by posts. These components need to be rot resistant and carefully sized for the load they'll carry. Posts and beams can be located and attached to each other in various ways. The location may be out at the edge of the deck or back several feet from the edge. Beams may be placed on the same plane as joists or underneath them. Posts usually go underneath beams, but they can be bolted to the sides of the beams. Long joists may require multiple beams.

**Sizing posts and beams**   Although posts are installed before beams, beams need to be located and tentatively sized in the design phase before you can determine post sizes. As discussed in Chapter 3, the number of posts needed on a deck is usually dictated by the beam size and span, although occasionally you may need to add extra posts due to poor soil conditions. The final choices may result from a series of considerations on size, span, cost and availability of materials.

Likewise, the size of the joists on a deck is determined by their maximum span and spacing, both of which are influenced by the load they carry and the decking thickness and pattern. Joist span can be cut in half by adding an extra beam, and smaller joists can save money. But an extra beam requires more foundation work, which will usually more than offset the savings on the joists. Sometimes, however, the deck is just too large or the longer joist size isn't available, leaving you with no choice but to add the second beam.

Beam size can be reduced by adding more beams, but it can also be done by supporting the beam with more posts. Adding an extra post or two will require extra foundation work too, but not as much as adding an extra line of posts needed to support a second beam. You cannot size one part of the frame without considering the changes it makes to other parts.

I prefer to use the minimum number of beams whenever possible, and I try to keep them no larger than a 4x12. Using one beam is almost always the most cost-effective choice, even if it dictates larger joists. Pressure-treated solid timbers are almost always available, and if they aren't it's easy to make a built-up beam. Finally, a 4x beam mates perfectly with the 4x4 posts that I prefer to use.

To select the right beam size, you need to know the species of wood you'll be using, the distance between beams and the number of posts. With that information, you can use the span table on the facing page to determine the beam size. Standard span tables tend to be conservative, but it's always better to overbuild than to underbuild. Your local building inspector should be able to tell you if this span table is suitable in your area.

To find the right size beam for a 12-ft. by 20-ft. deck with a single beam and ledger, look at the table under the column for 12-ft. beam spacing. This assumes that the beam will be placed at the end of the joists and therefore will carry exactly half the load (the ledger carries the other half). If you're using pressure-treated Douglas-fir for the beams and you're spacing the posts every 7 ft. (that is, the beam has to span 7 ft.), the table indicates that you should use a 4x10 beam. By spacing the posts every 6 ft., you could use a 3x10, which could be made using two 2x10s (when a nominal 3x beam is call for, I generally plan on a full 3 in. of wood, just to be on the safe side). This table is designed for multiple beams, so when you're using it on a single beam and ledger system the recommendations will be very conservative.

Since the beam is 20 ft. long, four posts would be required as long as the beam was a 4x10 or a 3x12. You could also use a 3x10 by moving up to five posts. I would choose among these options depending upon the availability of material, the cost and how high I had to lift the beam. The taller a beam is, the less stable it is sitting on a post. I'd prefer to use a shorter beam and to add an extra post, if it didn't require the posts to be spaced less than 6 ft. apart. Because of the conservatism of span tables, you might be able to use a beam that's smaller than what is called for and still be safe, but don't do it without consulting an engineer first.

If the posts at the ends of the beam are moved in a little, the distance between each post can be decreased slightly, possibly allowing for a smaller beam. For example, if the posts were moved in 1 ft. on each end of the 20-ft. beam above, the span from post to post would now be 6 ft., allowing the use of a 3x10 beam instead of a 3x12. I like to cantilever the beam past the posts in this manner for another reason—you don't have to worry about the foundation falling exactly at

## Typical Beam Spans

| Wood Species | Nominal Beam Size | Spacing Between Beams (Joist Span) in Feet | | | | | | | | |
| --- | --- | --- | --- | --- | --- | --- | --- | --- | --- | --- |
| | | 4 | 5 | 6 | 7 | 8 | 9 | 10 | 11 | 12 |
| | | Maximum Beam Span in Feet | | | | | | | | |
| Southern yellow pine, Douglas fir-larch | 4 x 6 | 6 | 6 | 6 | | | | | | |
| | 3 x 8 | 8 | 8 | 7 | 6 | 6 | 6 | | | |
| | 4 x 8 | 10 | 9 | 8 | 7 | 7 | 6 | 6 | 6 | |
| | 3 x 10 | 11 | 10 | 9 | 8 | 8 | 7 | 7 | 6 | 6 |
| | 4 x 10 | 12 | 11 | 10 | 9 | 9 | 8 | 8 | 7 | 7 |
| | 3 x 12 | | 12 | 11 | 10 | 9 | 9 | 8 | 8 | 8 |
| | 4 x 12 | | | 12 | 12 | 11 | 10 | 10 | 9 | 9 |
| | 6 x 10 | | | | | 12 | 11 | 10 | 10 | 10 |
| | 6 x 12 | | | | | | 12 | 12 | 12 | 12 |
| Hem-fir | 4 x 6 | 6 | 6 | | | | | | | |
| | 3 x 8 | 7 | 7 | 6 | 6 | | | | | |
| | 4 x 8 | 9 | 8 | 7 | 7 | 6 | 6 | | | |
| | 3 x 10 | 10 | 9 | 8 | 7 | 7 | 6 | 6 | 6 | |
| | 4 x 10 | 11 | 10 | 9 | 8 | 8 | 7 | 7 | 7 | 6 |
| | 3 x 12 | 12 | 11 | 10 | 9 | 8 | 8 | 7 | 7 | 7 |
| | 4 x 12 | | 12 | 11 | 10 | 10 | 9 | 9 | 8 | 8 |
| | 6 x 10 | | | 12 | 11 | 10 | 10 | 9 | 9 | 9 |
| | 6 x 12 | | | | 12 | 12 | 12 | 11 | 11 | 10 |
| Western red cedar, redwood | 4 x 6 | 6 | | | | | | | | |
| | 3 x 8 | 7 | 6 | | | | | | | |
| | 4 x 8 | 8 | 7 | 6 | 6 | | | | | |
| | 3 x 10 | 9 | 8 | 7 | 6 | 6 | 6 | | | |
| | 4 x 10 | 10 | 9 | 8 | 8 | 7 | 7 | 6 | 6 | 6 |
| | 3 x 12 | 11 | 10 | 9 | 8 | 7 | 7 | 7 | 6 | 6 |
| | 4 x 12 | 12 | 11 | 10 | 9 | 9 | 8 | 8 | 7 | 7 |
| | 6 x 10 | | 12 | 11 | 10 | 9 | 9 | 8 | 8 | 8 |
| | 6 x 12 | | | 12 | 12 | 11 | 11 | 10 | 10 | 8 |

*Note: All lumber should be #2 grade or better.*

the edge of the deck. As long as it is set back a bit, a discrepancy of a few inches won't be noticeable. This method won't work, however, if you're using continuous posts that are integral to the railing. My rule of thumb for cantilevering beams over posts is to cantilever no more than the height of the beam (see the drawing on p. 72).

Once you have a tentative beam size and number of posts, the next step is to look at the post-size table (see p. 82). As with the beam-span table, there are different standards for different woods. Sizing posts is based on the load they carry and on their length.

To use the table, multiply the beam spacing by the distance between posts to determine the square

## Typical Post Sizes

| Wood Species | Nominal Post Size | Load Area in Square Feet (beam spacing x post spacing) | | | | | | | | | |
| --- | --- | --- | --- | --- | --- | --- | --- | --- | --- | --- | --- |
| | | 36 | 48 | 60 | 72 | 84 | 96 | 108 | 120 | 132 | 144 |
| | | Maximum Post Height in Feet | | | | | | | | | |
| Southern yellow pine, Douglas fir-larch | 4 x 4 | 12 | 12 | 12 | 12 | 10 | 10 | 10 | 8 | 8 | 8 |
| | 4 x 6 | | | | | 12 | 12 | 12 | 12 | 10 | 10 |
| | 6 x 6 | | | | | | | | | 12 | 12 |
| Hem-fir | 4 x 4 | 12 | 12 | 10 | 10 | 10 | 8 | 8 | 8 | 8 | |
| | 4 x 6 | | | | 12 | 12 | 12 | 10 | 10 | 10 | 10 |
| | 6 x 6 | | | | 12 | 12 | 12 | 12 | 12 | 12 | 12 |
| Western red cedar, redwood | 4 x 4 | 12 | 10 | 10 | 8 | 8 | 8 | 6 | 6 | 6 | 6 |
| | 4 x 6 | | 12 | 12 | 10 | 10 | 10 | 8 | 8 | 8 | 8 |
| | 6 x 6 | | | | 12 | 12 | 12 | 12 | 12 | 12 | 12 |

Note: All lumber should be #2 grade or better.

footage supported by each post. Then look down the column to find the smallest post that will do the job for the height of the deck. It will be easiest to fasten posts and beams together if the post width is the same as the beam width. I always use square posts—they look better. If the posts are going to run through to become part of the railing, you need to size them with the railing in mind—a 6x6 would be too big for most railing designs. Whenever possible, I like to use 4x4 posts; 6x6s tend to check and crack more. But if you're using a 6x beam, you should also use a 6x post.

To size the posts in our 12-ft. by 20-ft. deck with four posts supporting the 20-ft. beam and no beam cantilever, the on-center spacing between posts is 6 ft. 8 in. Multiply this by the 12-ft. beam spacing, and the result is 80 sq. ft. With that figure, look under the 84-sq.-ft. column in the table, which indicates that you could use a 4x4 Douglas-fir post, provided it was no longer than 10 ft. tall.

**Beam types and locations** The beam can be in the same plane as the joists or it can be under the joists. If it's in the same plane, the joists will be hung on the beam with metal joist hangers, and the beam will normally rest on posts. In some situations, the beam will be bolted to the posts so that the post can continue up

through the deck to become part of the railing. With the beam and joists in the same plane, there will be more headroom under the deck, which might be necessary on a low deck. But this method of framing requires that the post bases on the foundation fall directly under an inflexibly located beam, which requires more precision. It's easier (and in my opinion, better looking) to put the beams under the joists, and to set the beam and the posts back from the edges of the deck. Creating different planes adds visual interest to a deck. Cantilevering joists over beams may allow you to use smaller joists by cutting down their spans. And this system allows for some margin of error in positioning the posts and beams.

Cantilevering the deck over the beam will change the load on the beam. The rule of thumb is that you can cantilever up to one-half the beam-to-beam spacing without changing the beam size. But I rarely find any reason to stretch these cantilevers to the maximum on a deck. The idea is to set the beam back from the front face a little for a better appearance and ease of installation, not to test your engineering skills. For 2x8 or 2x10 joists, I usually cantilever about 12 in. and seldom exceed 18 in.

Beams can be solid timbers or built-up of layers of pressure-treated 2x framing lumber. Compared with

## Post and Beam Systems

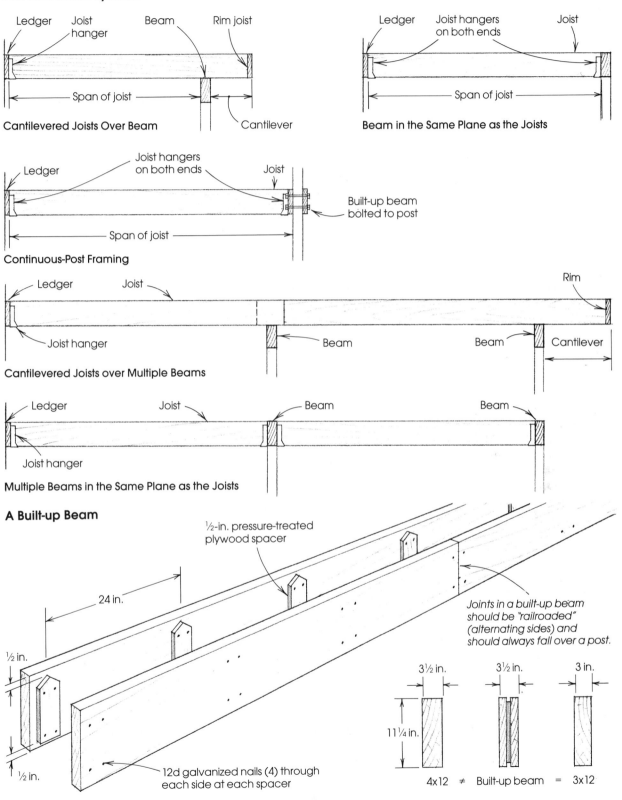

**Cantilevered Joists Over Beam**

Ledger · Joist hanger · Beam · Rim joist · Span of joist · Cantilever

**Beam in the Same Plane as the Joists**

Ledger · Joist hangers on both ends · Joist · Span of joist

**Continuous-Post Framing**

Ledger · Joist hangers on both ends · Joist · Built-up beam bolted to post · Span of joist

**Cantilevered Joists over Multiple Beams**

Ledger · Joist · Rim · Joist hanger · Beam · Beam · Cantilever

**Multiple Beams in the Same Plane as the Joists**

Ledger · Joist · Beam · Beam · Joist hanger

## A Built-up Beam

½-in. pressure-treated plywood spacer

24 in.

½ in.

½ in.

12d galvanized nails (4) through each side at each spacer

*Joints in a built-up beam should be "railroaded" (alternating sides) and should always fall over a post.*

3½ in.　3½ in.　3 in.

11¼ in.

4x12　≠　Built-up beam　=　3x12

built-up beams, solid timbers are usually more expensive, harder to find and not any stronger. All joints must fall over a post, which can create waste. But built-up beams do require extra labor. Timbers are usually designated 4x or 6x (their nominal size), but their dressed width is about ½ in. less. Span tables take this into account, so you can safely choose timbers by their nominal size.

Built-up beams are made by layering 2x framing lumber around spacers usually made with ½-in. pressure-treated plywood (see the drawing on p. 83). One of the most rot-prone places on a deck is where two boards are nailed together face to face. The water gets into the joint easily but can never really dry out. For best results, make sure that the spacers are pointed (diamond-shaped is good too), and place them about every 2 ft. in the built-up beam.

The spacers also have another benefit—a built-up beam composed of two 2x boards around ½-in. plywood creates a 3½-in. beam, which is the same width as a nominal 4x4 post. This makes connecting the two with standard metal post caps much easier than trying to fit a 3-in. solid beam on a 4x post.

The span table gives sizes for solid beams. A built-up beam made from two layers of 1½-in. material yields a total beam thickness of 3 in. Even if the boards are separated by ½-in. intermittent plywood spacers (making the finished beam 3½ in. thick), it's still only as strong as the two layers. It has been my experience that this built-up beam may be stronger than a solid 3-in. beam, perhaps because it spreads out the weaknesses in the wood. But when sizing beams from a span table, use the 3-in. column for a 3½-in. beam.

For example, if the span table calls for a 4x timber, a built-up beam of two 2x boards will not be its equivalent (3½ in. vs. 3 in.). You'll need to add a third layer to the built-up beam, even though it surpasses the strength required.

Joints in built-up beams should be "railroaded," that is, staggered so that they fall over posts, but with only one joint over any post. One board in the beam should always be continuous over a post. This takes a little planning and may waste some lumber, but it makes for a stronger beam.

**Installing posts**   It's quite possible that each post will need to be a different length on a deck. That's because the entire foundation is rarely at the same level. The easiest way to find the post length is to set them in place one at a time, mark them and then cut them (and brush some extra preservative on the cut end).

To determine post height, begin by marking a reference line below the ledger representing the far end of the joists. You should always slope the deck away from the house about 1/16 in. for every foot of joist length to drain water. For the 12-ft. by 24-ft. deck in our example, that would require a total drop of ¾ in. (12 x 1/16 = ¾). So place a mark ¾ in. below the ledger. If the beam is going to be in the same plane as the joists, this mark will indicate the desired height of the posts. If the beam is going to be under the joists, then drop this mark by the exact height of the beam.

Now set a post in place on its base. Using either a 4-ft. level on a straight board or a water level, transfer the reference mark below the ledger to the post. Then take the post down and cut it to length. Stand it back up and temporarily brace it in a plumb position with 1x4s nailed to the post and staked to the ground. You'll need to brace on two sides to keep the post plumb in each direction.

If the posts are going to be continuous up to the railing, then an additional reference mark needs to be placed above the ledger at a distance equal to the sum of the thickness of the decking and the height of the railing. Make sure that all the posts are long enough to reach a little beyond this height, then put them in place and temporarily brace them. Transfer the beam location to the posts as described above. The tops of the posts will be marked with a chalkline and cut to length after the decking is in place.

**Installing the beam**   When installing beams, you need to put the crown side up. Sight down along the edge of the beam to see which side arches up. If the crown is installed up, the edge will level itself under a load. If the crown is installed down, however, it will create a permanent sag in the deck frame.

The beam extends between the outside faces of the end joists. When positioning the beam, it may be necessary to reset the string lines that represent the edges of the deck. The beam should then reach from one string line to the other. Let the beam run a little long at the ends so that when the joists are installed and the framing squared up, there's a little room to move the joists to adjust for any small misplacement in the foundation and post system. Once the joists are installed, the beam can be cut flush with the sides of the deck and the bottom corner trimmed off for a neat look.

For the best support and ease of installation, I prefer to place beams on top of posts. This is an infallible approach that doesn't rely on bolts to hold the beam up, but it does require post caps at every intersection, and it may require hurricane ties at every joist. With a continuous-post system, however, the beam will need to be attached to one or both sides of the post with through bolts.

The height of the posts is determined by first establishing a reference mark on the house, which must take into account whether the beam is going to be below the joists or in the same plane. The reference should also incorporate the slope of the deck. If you're using a continuous-post system, make another reference mark higher on the house to indicate the top of the railing. Then transfer the reference mark or marks to the posts using a water level or a 4-ft. level resting on a long board.

Trim off the bottom corner of an exposed beam for a better looking deck.

**Locating Joints in Solid Beams**

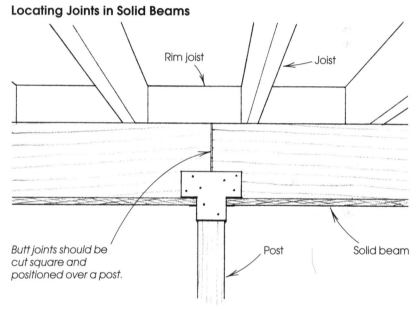

Rim joist

Joist

Butt joints should be cut square and positioned over a post.

Post

Solid beam

Lifting beams into place can be heavy work. If you have to lift a 20-ft. 4x10 beam over your head, you should count on needing four people to do it safely. I've often had to do it with just two carpenters by setting up scaffolding and moving the timber up in stages, one end at a time. Don't underestimate the weight of a beam this size or the damage it can do to a person if it falls. Make sure that all butt joints on solid beams are cut square and fall over a post. The joints should also be connected with two 1-in. by 18-in. metal straps.

A built-up beam is most easily built on the ground. Put down the first layer of 2x pieces, cut to length and in the order they'll go when it's finished, and take care to locate joints only over posts. Be sure to put the crowns of both 2x layers pointing in the same direction, and remember that the crowns will go up when they are installed. Then tack 6-in. wide plywood spacers every 2 ft. and over each joint. Place the other 2x layer in place, making sure to stagger the joints so that only one falls over any one post, and nail the beam together through the spacers (see the drawing on p. 83). Use four 12d galvanized nails at each spacer on each side. Work on a flat surface and keep the edges aligned. If the finished beam is going to be too long or heavy to lift into place, leave out a few nails at this point so that it can be lifted in sections. Finish the nailing once the beam is in place. Two ⅜-in. galvanized carriage bolts fastened through the beam at each spacer will add strength and may be a necessity if the boards are cupped.

As I've stated, I prefer to have beams resting entirely over posts when possible. But if the design calls for continuous posts, the beam will need to be attached to the sides of the posts. A built-up beam constructed with one layer of 2x stock on either side of the post can be made without notching the layers into the post—bolts are used for the connection, never nails alone. A larger solid timber placed on one side of a post could use bolts alone if they were large enough and numerous enough, but better support is ensured if the beam is notched into the post. Some builders I know prefer to notch and bolt all the beams to posts even on noncontinuous posts. Although this may increase lateral sway resistance, it's usually not enough to eliminate the need for diagonal bracing. I feel the gain isn't worth the extra labor involved in cutting and bolting.

With continuous posts, the beam is constructed and installed differently. If you're using a built-up beam, the layers of the beam should be bolted to the inside and outside of the posts to keep the load on each post centered (see the drawing below, left). Then use short pieces of post stock every 2 ft. in between the layers to act as spacers. Bolt each spacer with two ⅜-in. galvanized bolts. With the posts plumbed and secured by temporary braces, mark the beam heights on the posts as described previously, tack the layers of the beam to the posts with nails, and then drill holes for the bolts. If the posts are less than 8 ft. apart, I use two ½-in. galvanized bolts in each connection with beams up to 10 in. high. Use three bolts for 12-in. beams.

Holes for bolts should be close to the vertical centerline of the post, but offset horizontally at least one bolt diameter from each other. They should be placed about 1½ in. from the upper and lower edges of the beam. Drill holes the same diameter as the bolt through the post and beam.

Notching a beam into a post provides a shoulder for the support of the beam. Notching should be no deeper than 1 in. on a continuous 4x4 post and 2¾ in. on a 6x6 post. Obviously, using 6x6 posts will provide better support whenever notching is required. The number of bolts used is the same as described above. Whenever untreated wood is exposed in a notch, be sure to toughen it up with preservative.

To lay out the notches, set the posts in place, mark the beam locations (shown in the drawing below, right) and then lay the posts down to cut them. Cutting notches can be done after the post has been stood in position, albeit more dangerously. The first cuts are done at the lines marked for the top and bottom of the beam with the saw blade set at the depth of the notch.

## A Built-up Beam for a Continuous Post

## Notched Beams

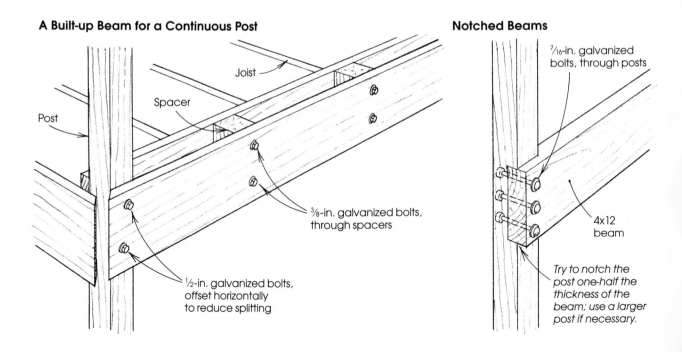

Joist

Spacer

Post

⅜-in. galvanized bolts, through spacers

½-in. galvanized bolts, offset horizontally to reduce splitting

⁷⁄₁₆-in. galvanized bolts, through posts

4x12 beam

*Try to notch the post one-half the thickness of the beam; use a larger post if necessary.*

After sawing the ends of the notch, make multiple passes with the saw, leaving ¼-in. wide wafers still in place. Knock these out with a hammer and finish the inside of the notch with a chisel.

**Post-to-beam connectors** Metal post caps (shown in the photo at right) are the most common fasteners for beams that sit on top of posts, and they should be fastened according to the manufacturer's instructions. They aren't very attractive, but you can paint them to match the deck. Before installing the post caps, make sure that the posts are plumb on all sides. Some people prefer to make a wooden gusset of 2x pressure-treated wood the same width as the post to lap the intersection and hang down the post about 18 in. Nailing on 1-in. by 18-in. metal straps will help tie the butt connections of timbers when using this method. The gusset should be beveled at both ends for drainage and a finished appearance, and it should be attached with four ⅜-in. lag bolts into both the post and the beam (see the drawing at right).

A good alternative is to use large, angled plywood gussets that measure 24 in. horizontally and lap down the post 18 in. These are not very attractive and so are usually used when they can be hidden (lower drawing at right).

**Bracing** There are no hard and fast rules for the amount of permanent bracing needed between posts and beams. Check with your local building department to find out what kind of bracing you'll need.

I like to add lateral sway bracing to all my decks, whether it's required or not, because it always makes the deck feel more solid. The simplest way to brace a deck is to bolt diagonal boards between the post and beam. They can be bolted to the backside to keep them less visible. The effectiveness of the bracing depends on how well it is fastened. Screws or small lag bolts are much better than nails. If the posts and beams are both solid wood, you might want to use 4x4 timbers as "knee" braces to give the structure a more attractive, timber-frame appearance. In this case, the braces should butt the post and beam and be attached with lag bolts (shown in the drawing on p. 88).

The taller the post, the longer the bracing needs to be. For posts under 4 ft., 2x4 diagonal braces are adequate. These 45° diagonal braces should come down 24 in. on the post. For an 8-ft. post, I use 2x6s that span at least 30 in. down the post. For posts over 8 ft. or in areas where seismic bracing is a must, a more complicated bracing system is required. When installing diagonal braces, I always make sure that the

A metal post cap ties the post to the beam. Post caps may be required on both sides.

### Site-Built Ties

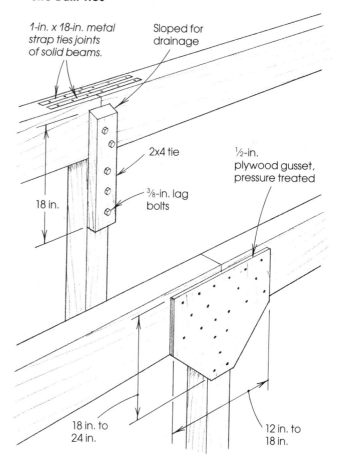

1-in. x 18-in. metal strap ties joints of solid beams.

Sloped for drainage

2x4 tie

⅜-in. lag bolts

18 in.

½-in. plywood gusset, pressure treated

18 in. to 24 in.

12 in. to 18 in.

Lateral sway bracing will stiffen any deck, whether or not it's required by code. It should be securely fastened to the post and the beam. Leave a small gap at the brace joint on the post to facilitate drainage.

## Solid Knee Braces

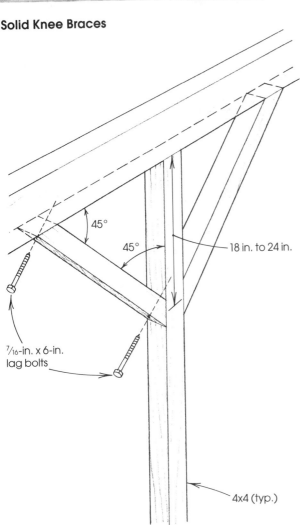

45°

45°

18 in. to 24 in.

$^{7}/_{16}$-in. x 6-in. lag bolts

4x4 (typ.)

end cuts are vertical, and I leave a space between the braces on the post to allow the water to drain through.

## Joists

On most decks, the joists are attached to the ledger on one end, extend away from the house and are capped with either a beam or with a single board called a rim joist. Freestanding decks will have rim joists or beams at both ends. Other parts of the system include headers, which are essentially small beams made of doubled joists that span a hole in the framing. The header supports any cut joists that would have gone through to the other side of the hole. Blocking is a term that refers to short pieces of perpendicular joist which go in between full-length joists.

Joists are prone to decay and should always be built of rot-resistant wood. Pressure-treated lumber is the best choice, although redwood and cedar are used on occasion for aesthetic reasons. Like posts and beams, joists are sized and spaced according to lumber type, span and load. The decking pattern may also affect the joist system. Joists are often a visible part of the framing, and when they are they should be constructed with care, using miters at the corners, for example. Alternatively, the perimeter of the joists can be covered with a layer of better-looking wood, called a fascia or skirt.

**Sizing joists** The object in sizing joists is to use the smallest and fewest joists possible to span the necessary distance. The decking material and pattern will often affect joist span and spacing. So the first thing to do is decide on the decking wood type, size and pattern. (See chapters 2 and 5 for more information on decking materials and patterns.) Next, look at the top table on p. 90 and find the recommended joist spacing for the decking you'll be using. Note that joists supporting diagonally laid decking must be closer together than joists that support perpendicular decking. Weaker or thinner deck boards also require closer joist spacing. The figures in this table are conservative and based on my experience; in some instances, the recommendations are even more conservative than those of the lumber associations, but I like to build a very solid deck. Once you determine the joist spacing, you can turn to the bottom table on p. 90. Note that the decking table determines the maximum distance between joists, but the span tables may require an even closer spacing. Spacing between joists is referred to as the "on-center" spacing, that is, the distance between centers of the joists; the joist span is the total length of the joist between supports.

For example, let's assume that on our 12-ft. by 20-ft. deck I want to lay 2x4 cedar decking perpendicular to the joists. The decking-span table (on p. 90) specifies the maximum distance between joists to be 16 in. By applying this information to the joist-span table, I see that if I want to use #2 Douglas-fir joists spaced 16 in. on center and spanning 12 ft., I'll need to use 2x8s. I'd also check if it was worth using smaller joists spaced closer together. In this case, however, 2x6 joists 12 in. on center can span only 10 ft. 9 in. At this point, I might want to ask myself if it would make sense to use 2x6 cedar decking, which costs about the same per square foot but requires joists spaced only 24 in. on center. The span table confirms that I could safely use 2x10 joists for a 12-ft. span. If I'm not too particular about 2x4 vs. 2x6 decking, I might make my final decision based solely on cost and availability.

There are a few other thoughts to keep in mind when sizing joists. Cantilevered joists technically should not hang out farther than one-fifth their maximum span for size and spacing, and joists should have twice as much length supported as the amount cantilevered to act as an anchor. Increasing the cantilever allows you to move the joist supports (ledger and beam) closer together, and this may permit you to use smaller joists. But longer cantilevers (even if they don't violate the one-fifth rule) can create a bouncier deck, and clients have a habit of noticing things like that. Also, using wider joists spaced farther apart will create a deck that feels more solid than narrower joists spaced closer together. As a practical minimum, 2x8s are the smallest joists I use.

One of the most rot-prone places on a deck is where the ends of two decking boards butt over a single joist. Try to avoid this problem by designing the joist system so that decking-board ends always fall over the

gap between a doubled joist with a ½-in. pressure-treated plywood spacer in between (see the photo below). Doubled joists cost more in material and labor, so it's wise to minimize them by choosing a decking pattern that requires only a few double joists. Double joists with plywood spacers will require 3½-in. joist hangers, which are commonly available.

**Joist layout** Before you can install any joists you need to mark their locations on the ledger and the rim or a beam. This is simple work, but it requires a consistent approach. Layout is usually marked on the rim joist when joists are cantilevered over the beam and the whole joist frame can be squared independently of the beam before attaching to it. Layout is marked on the beam when the joists are hung on a fixed beam. The starting location for the layout of joists must be precisely determined. The ledger should be laid out (that is, marked for joist position) only after it has been bolted to the wall. The rim joist is usually laid out before it is installed. The layout ensures that all the joists are installed parallel, at least until the framing changes levels or the decking changes direction. Start the layout from an edge or face that is unbroken and common to both ledger and rim joist. Then, by following the same spacing along the ledger and the rim joist, you'll be sure of an accurate layout. For most decks, this common edge is usually the outside face of one of the two end joists. On a more complicated structure with angled sides or offsets, the most practical common edge will be a joist somewhere in the middle of the frame.

**Design your deck so that decking joints fall over a gapped double joist. That way water will drain through.**

## Span Table for Decking

| | Joist Spacing in Inches (o. c.) | | |
|---|---|---|---|
| | 12 | 16 | 24 |
| Decking perpendicvular to joists | ¾ in. x 6-in. Southern yellow pine<br><br>5/4 x 4 or 5/4 x 6 redwood or cedar | 2 x 4 redwood or Western red cedar<br><br>5/4 x 4 or 5/4 x 6 Southern yellow pine or Douglas-fir | 2 x 6 or larger all species<br><br>2 x 4 Southern yellow pine or Douglas-fir |
| Decking diagonal to joists | 2 x 4 redwood or cedar<br><br>5/4 x 4 or 5/4 x 6 Southern yellow pine | 2 x 6 all species<br><br>2 x 4 Southern yellow pine or Douglas-fir | |

Note: All lumber should be #1 grade or better.

## Span Table for Joists

| | Joist Size | | | | | | | | | | | |
|---|---|---|---|---|---|---|---|---|---|---|---|---|
| | 2x6 | | | 2x8 | | | 2x10 | | | 2x12 | | |
| | 12 in. o. c. | 16 in. o. c. | 24 in. o. c. | 12 in. o. c. | 16 in. o. c. | 24 in. o. c. | 12 in. o. c. | 16 in. o. c. | 24 in. o. c. | 12 in. o. c. | 16 in. o. c. | 24 in. o. c. |
| Wood Species | Maximum Span | | | | | | | | | | | |
| Douglas fir-larch | 10 ft. 9 in. | 9 ft. 9 in. | 8 ft. 1 in. | 14 ft. 2 in. | 12 ft. 7 in. | 10 ft. 3 in. | 17 ft. 9 in. | 15 ft. 5 in. | 12 ft. 7 in. | 20 ft. 7 in. | 17 ft. 10 in. | 14 ft. 7 in. |
| Hem-fir | 10 ft. 0 in | 9 ft. 1 in. | 7 ft. 11 in. | 13 ft. 2 in. | 12 ft. 0 in. | 10 ft. 2 in. | 16 ft. 10 in. | 15 ft. 2 in. | 12 ft. 5 in. | 20 ft. 4 in. | 17 ft. 7 in. | 14 ft. 4 in. |
| Southern yellow pine | 10 ft. 9 in. | 9 ft. 9 in. | 8 ft. 6 in. | 14 ft. 2 in. | 12 ft. 10 in. | 11 ft. 0 in. | 18 ft. 0 in. | 16 ft. 1 in. | 13 ft. 2 in. | 21 ft. 9 in. | 18 ft. 10 in. | 15 ft. 4 in. |
| Redwood | NA | 7 ft. 3 in. | 6 ft. 0 in. | NA | 10 ft. 9 in. | 8 ft. 9 in. | NA | 13 ft. 6 in. | 11 ft. 0 in. | | | |

Note: All lumber should be #2 grade or better. Redwood should be construction grade or better.

Start the layout by pulling out several feet of a tape measure and laying the tape on the top edge of the ledger. Since the end joist isn't in place yet, you need to extend the tape over the edge of the ledger an extra 1½ in. Then mark along the top of the ledger to indicate the on-center locations of the joist (usually every 16 in. or 24 in.). I like to lay out for the side of the joist rather than the center, which can be done by adjusting the tape or by subtracting ¾ in. from each layout mark. Then I go back with a square and put a vertical line on the face of the ledger, which indicates the edge of each joist. This makes it easier to get the joist hangers on straight. Don't worry if the last pair of joists on the layout is spaced closer than the others.

To lay out the rim joist, first cut it to length. On a rectangular deck, the length will be the same as the width of the deck. If the rim joist is going to be exposed, it should be mitered on the ends. Start the layout from the edge of the rim joist—you don't need to compensate for the edge joist as you did with the ledger.

Decks are not always simple rectangles. Sometimes they have jogs or turns in their perimeter that result in ledgers or rim joists that are not continuous. When the layout for the ledger and rim start from a common edge, continuation of the layout at these offsets is simply a matter of using an edge of a joist common to both parts as a reference line to maintain the same relative spacings.

Similarly, if both the ledger and rim joist don't start at the same outside face of the deck, you need to choose a different common reference. For example, you may lay out the entire ledger starting at the end as described earlier, and then choose the edge of a middle joist (instead of the end joist) as the starting pint from which to lay out the rim joist. I suggest that you make sketches of these complicated layouts to show dimensions and indicate relative positions before starting the actual layout.

On some decks with cantilevered joists, I run the outside layout on the beam rather than the rim joist. I did it this on the Morrises' deck simply because it was easier to reach the beam than the rim joist. After the ledger was laid out, I established a common edge by nailing a full-length, straight joist in place, perfectly square to the ledger. Then I measured off of the joist, marking my layout on the beam. Joist framing can still be squared before attaching it to the beam.

The layout for the far end of the joists that butt into a beam (where the posts are continuous to above-deck levels, for instance) requires that the starting point on the beam be located precisely at the beginning, because joists cannot be adjusted afterward as they can in a cantilevered situation. String lines that marked the sides of the deck framing on the beam must be reinstalled. Once they're marked on the beam, it's a good idea to double-check the four outside corners of the deck at this point by measuring diagonals (shown in the drawing on p. 61) and make sure that everything is still square. If the beam ends were left long you'll have some latitude to shift things if necessary.

**Installing joists**  With the layout complete, the joists can be installed, beginning at the ledger. Some books on framing and deck building suggest that joists be installed using a secondary ledger, which is a 2x2 strip of wood. This is an old technique that requires a small ledger to be fastened to the main ledger. Then the joists are notched on the bottom and placed on the secondary ledger. The whole structure is then secured with a good dose of toenails. I don't recommend this method because the carrying capacity of the joists is reduced and the likelihood of splitting the joists is increased. It also adds an extra surface for water to collect on.

I prefer to support joists on the ledger with metal joist hangers. If the supporting beam at the other end is in the same plane as the joists, joist hangers can be used there as well. To install joist hangers, I use a small scrap of joist stock as a guide, holding it on the layout line and flush with the top of the ledger (see the top photo on the facing page). After putting in two nails, I remove the block and finish nailing.

Normal joist hangers work only when the joist meets the ledger or rim joist at a right angle. If the joist needs to join at something other than 90°, there are several options. Simpson Strong-Tie (see the Resource Guide on p. 150) makes joist hangers for 45° applications as well as adjustable ones that work for a range of angles. I usually have to special order these hangers, which isn't worth the bother and expense. Instead, I cut the joist to the proper angle and then nail it on with 12d or 16d galvanized nails. For added strength with a long joist, I use a bendable metal plate (the Simpson LS) to tie the two boards together.

With the joist hangers in place, the joists are cut to length, set in place and nailed to the hangers. The joists should be set square to the ledger, which is easy to do using the "Schuttner Square" described in Chapter 3, and installed with the crowns up (as explained on p. 84).

If the joists are going in hangers at the ledger and beam ends, they'll have to be individually measured and cut to length first. When the joists cantilever over the beam, however, don't bother cutting them to length until after they've been installed. That way you can snap a straight chalkline across all the joists to guide your cuts and ensure a straight line.

Joist hangers can be aligned by using a scrap piece of joist stock as a guide. Place the guide in the joist hanger, align the top with the top of the beam and then nail on the hanger.

Joists should be installed square with the ledger. Square up the joist using a "Schuttner Square" (see p. 62) and then temporarily tack it in place—it will be permanently fastened later on.

On most of the decks I build, the outside joists lap over the ends of the ledger. This serves the dual purpose of hiding the ends of the ledger (which looks better) and providing a nailing surface for securing the end joist. The end joists are nailed to the ledger with 16d galvanized nails. Take care not to split the ends (predrilling the nail holes is wise). For additional support, I use a 90° metal clip called a framing anchor to reinforce the connection on the back side (a normal joist hanger can be clipped in half and used too).

If the joists are going to be exposed on the finished deck, the outside corners should be mitered for the best appearance. Miters are most commonly used where the end joists meet the rim joist or a beam, but they can also be used at 45° or 90° turns in the rim joist. Mitered joists are longer than those that butt into the rim joist, so don't forget to save a few long boards for this purpose. Mitered corners should be joined with nails from both directions. Keep them back as far as possible from the corner to avoid splits.

**Lapping joists** Large decks may require a second set of joists and an intermediate beam. If the beams are in the same plane as the joists, then each set of joists will rest in its own set of joist hangers. If the beams are underneath the joists, then the joists will lap each other over the intermediate beam. The joists should overlap at least 12 in., and both ends should overhang the beam. To ensure good drainage, the joists should be separated by ½-in. plywood spacers and well nailed with five 16d galvanized nails. Lapping joists, although an easy and secure way to accomplish a transition, create problems on the layout. As shown in the drawing on the facing page, the joist centers on the lapping joists need to be shifted 2 in. Account for this shift on the rim-joist layout by placing the end of the tape 2 in. beyond the end of the rim. For the best appearance, don't lap the end joists. Instead, butt them into each other and splice them together on the inside with a short section of joist stock. Before securing them to the beam, use the layout strings to make sure these butted joists are straight.

**The rim joist** With cantilevered joists, the rim is given the same layout as the ledger. The straighter the rim stock is, the easier it will be to install. With a helper holding one end of the rim flush with the tops of the joists, drive several nails through the rim and into the end joist on the other side. Then move to the next joist, making sure that the tops of the joists are flush. Continue along the rim until each of the joists has been nailed with three or four equally spaced 16d galvanized nails. When I'm nailing on the rim joist, I first

## Lapping Joists

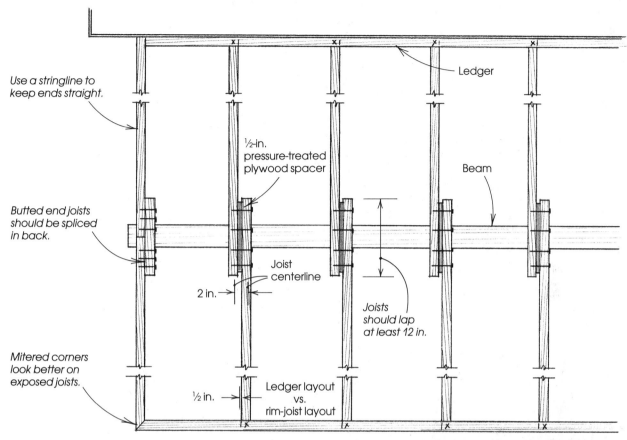

*Use a stringline to keep ends straight.*

Ledger

*Butted end joists should be spliced in back.*

½-in. pressure-treated plywood spacer

Beam

Joist centerline

2 in.

*Joists should lap at least 12 in.*

*Mitered corners look better on exposed joists.*

½ in.

Ledger layout vs. rim-joist layout

drive the top nail home, then use a small framing square to align the joist before nailing the bottom. If there are any joints in the rim joist, they must fall at the end of a joist. Miter the ends of the joint if it will be exposed.

If you marked a joist layout on the beam as described earlier, you won't need to lay out the rim joist. Just cut the joists to length, if they haven't been already, and nail on the rim joist as described above.

**Squaring and securing the frame** With the basic outline of the deck now defined by the joists, you need to check that the joists are sitting square with the ledger before securing them to the beam. This can be done by checking the diagonals. If they're within ⅛ in. of each other for each 8 ft. of length of the diagonal, the frame is square. Any deviation larger than this will require that you shift the frame one way or the other. This may require some persuasion with a large hammer, but don't be too brutal. As long as you left the

To fasten the lapping end joist to the ledger, first nail through the outside face of the joist into the ledger, then attach a metal clip to the inside corner. A standard joist hanger cut in half is perfect for the job.

beams a little long, you should be able to square up the joists without sliding off the beam on one end.

Now the joists can be fastened permanently to the beam. Some codes require that joists be attached to the beam with hurricane ties (see p. 34). I like to use them even when they aren't required—they're easy to install, and they add strength to the deck. For best results, put the anchors on alternating sides of the beam, one per joist. Tall and narrow beams, such as a 3x12, should have anchors on both sides of every joist. The anchors require special 1½-in. nails. Put one in each hole in the anchor. The option is to toenail the joists to the beam. Toenailing is frequently done and is generally permitted, but it's effective only if you don't split the joist in the process. Toenail with one 10d galvanized nail on each side.

Although generally viewed as questionable in value, blocking between joists is required by some codes to add lateral rigidity to the deck and spread the load more evenly. Blocking is generally needed only on joists that are longer than 12 ft., with one row needed every 8 ft. There is no quick way to install it. Snap a chalkline along the top of the joists, and then put a block (cut from joist stock) on one side of the line and then the other. Staggering the blocking in this manner makes it easier to nail through the joist. A few nails into each end of the block will do. Make sure that the blocking is not cut too long or the accumulated effect

of the difference will quickly throw the joists off their layout. Check the joists for straightness frequently.

If you're planning to add a fascia to cover the joists, now might be a good time to do the work if the fascia won't be covering the ends of the decking as well as the joists. See Chapter 5 for more on installing fascia.

## Special Framing Techniques

This section discusses some framing techniques you can use if you want to build a deck that's more than a simple flat box. Some of these special features require only modifying joist positions, while others must be tackled from the foundation on. To minimize extra work and building errors, all of these decisions should be made at the design stage (see Chapter 1).

**Turning corners**  One of the biggest decisions you have to make on a deck that wraps around to a second side of the house is how to handle the decking pattern. From there you need to design the joisting pattern. The least attractive option is for the decking to run continuously from one end to the other, with the decking around the corner meeting it at 90° along the house line (see the drawing at left on the facing page). In this case, the joists and beams would simply extend out to the end of the deck to support the decking. The

**With a helper to keep the rim aligned with the joists, drive one nail into each joist near the top of the rim joist. Then use a small framing square to align the joist (above) before driving the bottom nail home.**

## Turning Corners: Decking Patterns

### An Uninspiring Decking Pattern

### Mitered Decking Creates a More Appealing Pattern

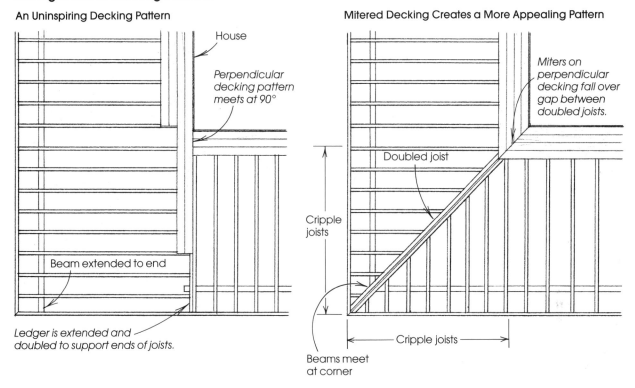

House

*Perpendicular decking pattern meets at 90°*

Beam extended to end

*Ledger is extended and doubled to support ends of joists.*

*Miters on perpendicular decking fall over gap between doubled joists.*

Doubled joist

Cripple joists

Cripple joists

Beams meet at corner

### Special Framing for Decking Boards Running in the Same Direction

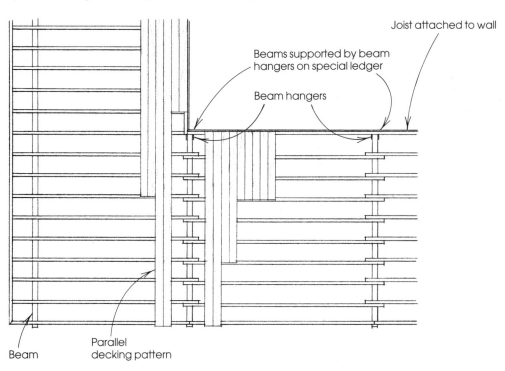

Joist attached to wall

Beams supported by beam hangers on special ledger

Beam hangers

Beam

Parallel decking pattern

ledger would have to be extended or a short beam used to carry the joists at the corner. A second set of joists perpendicular to the wall with a beam continues the decking around the corner.

A more attractive way to turn the corner with the decking is to create a mitered corner, with the ends of the mitered decking boards meeting over a doubled joist (drawing at right, p. 95). The joists run perpendicular to the wall on both sides of the corner, except for the doubled corner joist, which extends diagonally from the corner of the house. Cripple joists fill in the corners, running from the rims to the double joist. The two beams run parallel to the house and meet at the corner.

A third option is to keep the decking running in the same direction on the whole deck (drawing at bottom, p. 95). That is, the decking would be parallel with the house on one side and perpendicular on the other. The joists for the entire deck would be parallel on this deck, and this would require a change in tactics for joist installation. On one side of the house, the joists would no longer hang from the ledger. Instead, they would run parallel to the ledger, and the ledger would now technically be considered a joist itself. And the beams on this side of the house would be perpendicular to the house. These beams could be supported at the house line by their own ledger, or they could be fully supported on individual foundations. Beam sizing and post spacing specifications would be the same for all the beams as long as joist spans didn't increase.

This type of joist arrangement can be used any time the joists are parallel to the wall, not just when turning a corner (for example, if the decking were to be perpendicular to the wall of the house instead of parallel). It means a minimum of two beams, which requires extra foundation work to support both ends of the joists. However, it does eliminate a ledger for supporting the joists, but if the deck is to be connected to the house for stability, the end joist will need to be attached in the same way as a ledger with all its attendant flashing precautions (see p. 78 for more on flashing).

**Level changes**   Building on two or more levels is a simple way to add a special touch to an otherwise humdrum deck, and it's sometimes necessary to include level changes on particularly tricky sites. The difference in elevation between levels should be designed to allow for a comfortable rise to each step connecting levels—I think the rise should be less than that used for interior stairs. It's easiest to design level changes to rise and fall by the height of the joists to eliminate a

## Framing Level Changes

### A Small Deck on Top of a Larger Deck

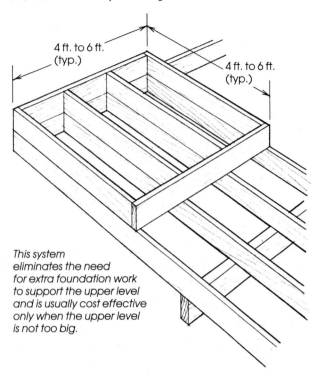

4 ft. to 6 ft. (typ.)

4 ft. to 6 ft. (typ.)

*This system eliminates the need for extra foundation work to support the upper level and is usually cost effective only when the upper level is not too big.*

### Separate Levels Supported Independently

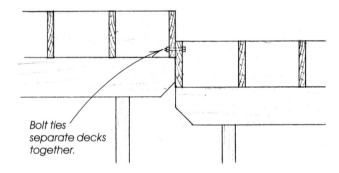

*Bolt ties separate decks together.*

### Separate Levels Supported by One Beam

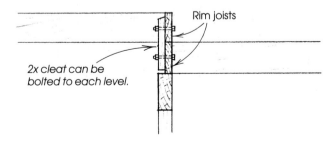

Rim joists

*2x cleat can be bolted to each level.*

This upper-level beam laps over and is supported by the lower level. The upper beam is notched to position it at the desired height. The ends of both decks are now supported by a single post.

An intermediate step connects the upper and lower levels and helps tie the two together structurally and visually.

An upper level with a large drop to the lower level is supported with small 2x8 posts off the lower beam. The levels will be joined with wide stairs.

A small 45° angle can be framed on a corner by cutting both joists an equal distance from the corner and then filling in with a diagonal. All cuts should be at 22½°.

A longer angled corner will require that the beam follow the angle as well.

lot of notching and shimming. (For more information on building stairs, see Chapter 7).

There are several ways to accomplish level changes. The easiest is simply to build another, smaller deck entirely on top of the first (see the top drawing on p. 96). This is really only practical if the upper level is relatively small—it saves on foundation work but requires some duplication of joists. This technique is common on landings smaller than 5 ft. or 6 ft. in front of doors.

A second way to accomplish level changes is to support each level independently (center drawing, p. 96). Each level would have its own foundation and framing, and if you need to make one of the levels freestanding, the levels should be linked with some bolts or extra framing.

A third option (bottom drawing, p. 96) requires the upper level to overlap and rest on a part of the lower level. Both levels are supported by the beams and posts of the lower level. The exact configuration of the framing will differ from one deck to another using this system, depending upon the differences in elevation and the direction of the joists and beams. Be sure that the lower-level joists are adequately supported themselves. Upper levels must have adequate bearing on lower levels, and both levels should to fastened together with anchors and nails.

**Angles** It's easy to break up the boxy look of a deck by putting 45° angles on the outside corners. On a square corner of a deck where the joists cantilever over the beams, small 45s can be built with simple joist changes. The amount of the cantilever will determine the maximum length of the angled piece. Larger 45s almost always require planning right from the design stage, as beams and supporting posts will need to be located to support the changes in joist framing. The small 45° angle is laid out by measuring equal distances from the corner along the outside upper edge of the rim and the end joist. Mark the two spots and then connect them with a chalkline if there are any intermediate joists. The rim and end joists should then be cut at a 22½° angle (half of a 45° angle), and any intermediate joists will need to be cut at a 45° angle 1½ in. back from the chalkline. Then measure the distance between the outside faces of the rim and the end joist and cut a piece of joist stock to fill in the diagonal.

Larger 45° corners require that the beam and its supporting posts also make a 45° turn. The beam and joist sizes will rarely need to be increased since the joist spans are being decreased. The most common framing designs are shown on the drawings on the facing page. The procedures described below will work if the beam is in the same plane as the joists or if it is under the joists.

## Turning a Corner with a 45° Angle

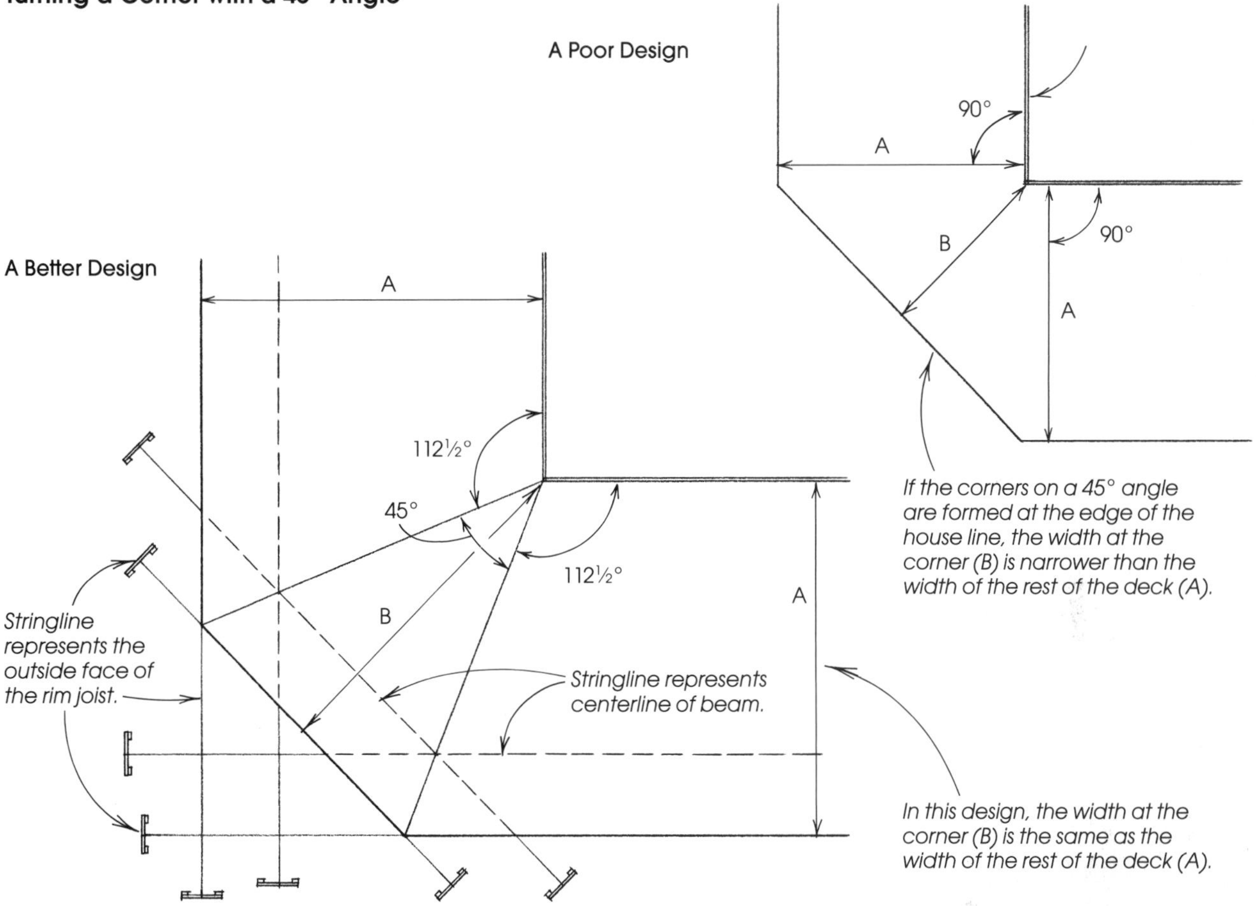

### A Poor Design

90°

A

90°

B

A

*If the corners on a 45° angle are formed at the edge of the house line, the width at the corner (B) is narrower than the width of the rest of the deck (A).*

### A Better Design

A

112½°

45°

112½°

B

A

*Stringline represents the outside face of the rim joist.*

*Stringline represents centerline of beam.*

*In this design, the width at the corner (B) is the same as the width of the rest of the deck (A).*

When I put a 45° corner on a deck that goes around the corner of a building, I like to keep the deck the same width all the way around. If the diagonal is placed as shown in the top right drawing above, the width of the deck will decrease. To keep the width constant, the diagonal needs to be the same distance from the corner as the width of the rest of the deck, which is shown in the left drawing. I use strings to represent these differences when I'm installing the foundation.

The foundation layout begins by installing batter boards and strings as for a 90° corner (see Chapter 3). Then measure back from the corner along the strings the desired distance. Stretch another string across the diagonal to represent the outside face of the 45. Then measure back from the diagonal to locate the centerline for the corner beam (see the drawing above). I try to keep the cantilevered distances the same through-

out the whole deck. Then add posts to hold the ends of the beam.

Joist framing at the corner begins with doubled joists radiating out from the building corner across the points where the beam changes directions. Then fill in with rim, cripples and any full-length joists. Keep them oriented according to the decking pattern desired, which usually means perpendicular to the rim joist.

If you're adding a larger diagonal that cuts across several intermediate joists to a deck that's not going around a corner of a building, the joists will remain parallel (the decking isn't changing directions). After installing extra posts and beams parallel to the diagonal as described above, it doesn't make sense to build a square corner of joists first and then cut off many feet of excess joist. Instead, install a rim joist and an end joist that are long enough to reach the diagonal, but leave out the intermediate joists for now. Cut the

## Framing a Curved Corner

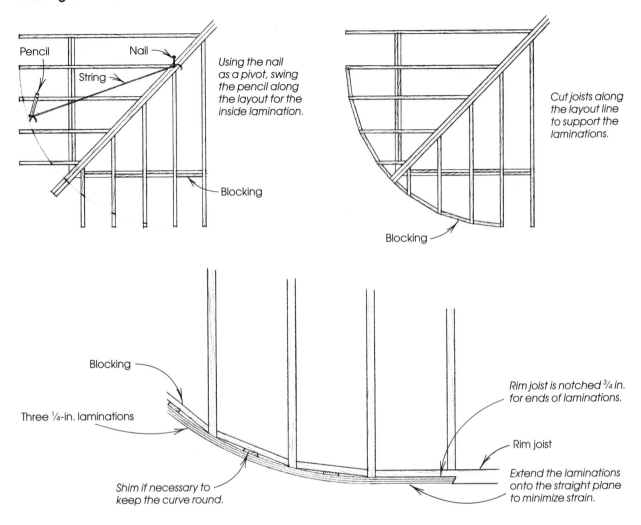

Pencil

Nail

String

*Using the nail as a pivot, swing the pencil along the layout for the inside lamination.*

Blocking

*Cut joists along the layout line to support the laminations.*

Blocking

Blocking

Three ¼-in. laminations

*Rim joist is notched ¾ in. for ends of laminations.*

Rim joist

*Shim if necessary to keep the curve round.*

*Extend the laminations onto the straight plane to minimize strain.*

miters on the rim and end joists, and then measure, cut and install the diagonal rim. The final step is to lay out and install the intermediate joists.

**Curves** Curves can place a deck in a class of its own. Unfortunately, curves are the most expensive shape to construct. Rim joists and the fascia covering the rim joists are the most frequently curved part of a deck frame, and they're typically built with glued laminations. In this process, thin layers of wood with glue in between are bent around a form. Because they're so large, beams usually aren't curved. Rather, the beam supporting a curved corner is installed at a 45° angle. Curved framing, in turn, requires curved railings, which are discussed on p. 130.

Wood for laminating needs to be clear and free from defects. The grain of the wood should run as parallel to the long dimension as possible. These qualities aren't usually available in pressure-treated wood, and so most curves are built of redwood or cedar. Other clear lumber can be used if it is thoroughly treated with a preservative after gluing.

For a curve 3 ft. in diameter, the individual layers may need to be as thin as ⅛ in. For curves of 10 ft. in diameter, ⅜-in. to ½-in. layers should work. All woods bend differently, and even different boards from the same order may bend differently. Green lumber will bend a lot more easily than kiln-dried stock. It takes many layers to build up a curved lamination, so it may be worthwhile to reduce the thickness of the rim joist to ¾ in. or 1 in. The reduced thickness won't be apparent provided it is flush with the outside face of the rim and covered with decking. When a curve is flush with the top of the decking, its thickness should be consistent with the rest of the fascia. At some point a transition will be made between the laminated curve and the straight section of unlaminated rim or skirt. The connection will be the smoothest if the lamination can straighten out for a foot or two before meeting the straight section. It's best to use continuous pieces of wood for the layers in the laminated section wherever possible because butt joints, even when staggered, are troublesome in a curve. They don't like to conform to the desired shape.

It's important that you use a good waterproof glue for the laminations. White glue and most yellow glues aren't waterproof and won't hold up to either the weather or the strains of the bent wood. Regular powdered, plastic-resin glues, although highly water resistant, are also not adequate. At the very least, you'll need a marine grade of plastic-resin glue, which needs an outdoor temperature of 70°F or warmer to set. Epoxies are another possibility, and they can often be used in cooler weather. Resorcinol is a good choice, and it's readily available, but it also requires an outdoor temperature of 70°F or higher. Whatever you use, you'll use a lot, so check prices before deciding.

To bend wood, you need a form. You can make the framing double as a form by applying the laminations directly to the framing. Once the framing is laid out

for the curve, it should be cut and blocked firmly into place. Then start applying layers of wood, using glue, fasteners and clamps until you reach the outer surface.

As an example, here's how I put a curve into the corner of a deck. First, I maximize the number of joists that contact the curve—the more joists you can hit, the smoother the form will be (see the drawing on the facing page). Install the joists as though you were building a square corner, leaving them long, and then put a row of blocking over the beam between the joists. The blocking will keep the joists locked in place while laying out the curve. For the arc you want to lay out, find the center of its circle by measuring back from both rims the radius of the circle, and use a nail to mark the center. Attach a string the length of the radius to the nail and mark the tops of the joists by drawing along the arc with a pencil. Now cut each joist at the appropriate angle, and again block between them, flush with the cut, to hold them exactly vertical.

Put the first layer of lamination in place and fasten it to the end of each joist with galvanized screws. The heads of the screws must be countersunk to ensure good fit between layers. Avoid flat spots by adding shims or small blocks between the first lamination and the blocking installed between the joists.

The remaining laminations are applied using glue and screws, nails or pneumatic staples and probably some clamps to hold the layers together until the glue sets. These will be covered by succeeding laminations,

## Framing for a Hole in the Decking

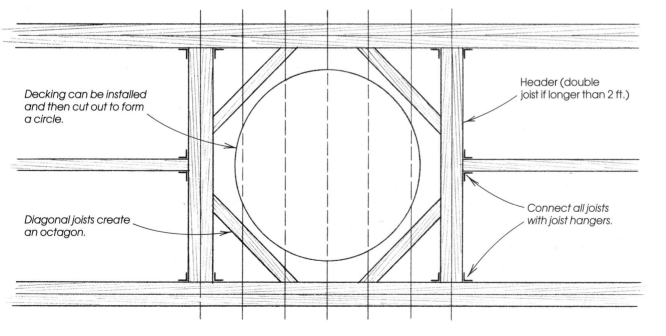

Decking can be installed and then cut out to form a circle.

Diagonal joists create an octagon.

Header (double joist if longer than 2 ft.)

Connect all joists with joist hangers.

so their appearance isn't critical. The fasteners in the last lamination should look neat, as they will be exposed. Fasteners can be eliminated on the last layer if you use glue and enough clamps to avoid gaps. After the glue has set, remove the clamps and clean up the top and bottom edges with a plane or belt sander.

Laminating wood is very rewarding, but it can be a frustrating and messy experience until you get the hang of it. Unless you want to buy overly wide lumber and trim it later, you need to work especially hard to keep the top edges aligned—a helper really comes in handy for this. To avoid gaps between layers, which look bad and let in water, use plenty of clamps and glue.

Install decking so that it overhangs the lamination and hides the glue joints, and be sure to nail the decking to the blocking between joists and not into the laminations.

If you want to put in a curved fascia to cover the ends of the decking boards, it should go on last. Install the decking and trim it to conform with the curve. This way the decking becomes a continuous component of the form for attaching the fascia.

**Framing around trees and hot tubs**  Occasionally a deck design will need to incorporate a tree, hot tub or other irregular object. If you need to frame around a round object, it looks best to make the hole in the decking round as well. Fortunately, this doesn't require that the framing be round. Framing for such obstacles is just a matter of leaving a square hole in the regular deck and "heading off" any joists that are interrupted (as shown on p. 101). Then, place short diagonal joists across the corners of the square to create an octagon. When laying the decking, allow the deck boards to cantilever over the joists and cut them to the desired shape around the object. Two-by-six decking can cantilever 8 in. or more without a problem. Leave a couple of inches of space around a tree so that it won't rub the deck when it sways in the wind, and don't make the joist framing too tight; allow for future growth.

In addition to an opening in the decking, a hot tub must have its own foundation, which is frequently a concrete slab. If the hot tub is going to sit on the deck rather than poke through it, you'll need to beef up the support structure for that area of the deck. A hot tub can easily exceed a load of 100 psf, which is a lot more than normal framing allows for. Manufacturers specify the structural requirements needed to support their tubs, so check. A good supplier will have suggestions on how to meet these specifications. If in doubt, seek some engineering help.

Hot tubs and spas need to have access hatches for pumps and filters. Electrical codes are very stringent about wiring requirements, so be sure to check these with the supplier and your local building inspectors.

## Building a Deck Over a Concrete Patio

For many people today, decks serve the same function that patios did a few decades ago. If you have a concrete patio, you may be able to build a deck right on top of it, using the patio itself as the foundation. You should do this only if the patio is in good shape and not extensively broken, misaligned or settled. Patios of brick or stone, unless built on a concrete slab, are probably not stable enough to serve as a base. Lay pressure-treated 2x4 "sleepers" on the patio (see the drawing on the facing page), with the wide side laid flat, to act as joists. These sleepers, like joists on a regular deck, should be spaced according to the type of decking used and its pattern. Nail the sleepers to the slab using a rented powder-actuated nailer. Hand nailing won't work in old concrete. Use pressure-treated wood shims under any low spots. If the patio is pitched much more than the amount I recommend for decks, as is likely, you'll need to do a lot of shimming (scraps of asphalt shingles work well for thin shims). When using wedge-shaped shims, it's best to put one in from each side—that way the board will be well balanced.

If you want to cover the patio completely, put sleepers all around the perimeter, flush with the edges of the patio, leaving small breaks to allow for drainage. The concrete edges and sleepers can be covered with a slim fascia. Put the decking down in the normal manner.

If there's enough vertical room, it'll be easier to use pressure-treated joists on edge and build a regular deck over the patio, complete with rim joist, ledger, and intermediate joists. The framing can be locked in place with blocks that are fastened with a powder-actuated nailer to the concrete and attached to the joists with metal clips. The joists, which could be as small as 2x6, may need to be shimmed. Technically, you could get by using 2x4 joists, but 2x4s on edge will require too many shims and are often too crooked for a flat finished surface.

## Building a Deck on a Roof

Building a deck on a flat roof isn't necessarily difficult, but a large roof that doubles as a deck should be designed for this dual function from the beginning. It needs to be capable of supporting all extra framing, people, furniture and snow loads. Since you'll be adding at least an additional 10 psf load to the roof, I

## Building a Deck over a Patio

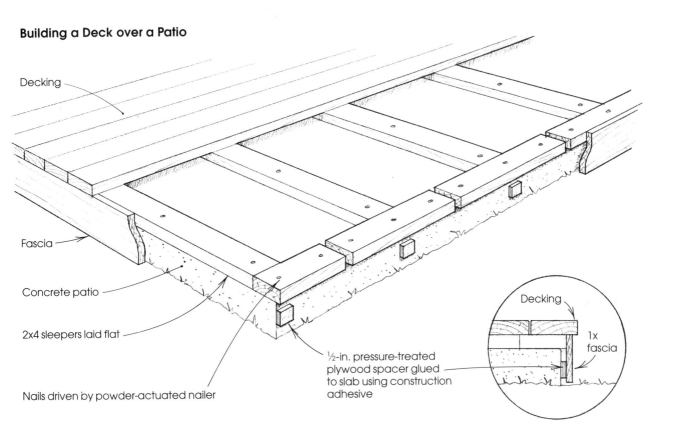

Decking

Fascia

Concrete patio

2x4 sleepers laid flat

Nails driven by powder-actuated nailer

½-in. pressure-treated plywood spacer glued to slab using construction adhesive

Decking

1x fascia

recommend that you seek the services of an architect or engineer, who can help determine if the existing roof is structurally adequate and can help you design a new one if it isn't .

Before you can put a deck on the roof, you need to make sure that you have a waterproof roof. Use closely spaced rafters and a minimum of ¾-in. tongue-and-groove plywood sheathing for such a roof. It's a good idea for rafters and substrate to slope ¼ in. per foot for drainage, even if the manufacturer of the waterproof roofing material states that it is suitable for a perfectly flat roof. There are many different types of water-proofing material available for low-pitched roofs, ranging from traditional built-up bituminous coating to the new generation of single-ply membranes.

Most single-ply membranes are large, loose skins that can be held in place by the weight of the deck. Examples of these synthetic rubber type materials are EPDM (ethylene propylene diene monomer) and CSPE (chlorosulphated polyethylene, commonly known by the trade name Hypalon). Modified bitumen is a tough, single-layer roll-type product that is heat sealed to the roof sheathing. The common bituminous multi-ply is "glued" down with hot tar and gravel. While

most flat roofs will benefit from the use of a single-ply roofing membranes, some types will work better under a deck than others. (For a good primer on the subject, see *Fine Homebuilding* magazine, No. 64, p. 43.)

Most roofs can be decked over once the waterproofing material is installed according to the manufacturer's recommendations. The big question to settle is what kind of deck to build. The secret of success is to build a deck without putting any holes in the roof.

One way to do this is to build a regular deck over the roof, but with pressure-treated joists on edge that have been ripped to a taper. If the taper matches the pitch of the roof, you can be assured of a level deck. A ledger can be bolted to the wall as described earlier in this chapter, and the joists can be hung on it with joist hangers. The joists don't have to be large as long as they align with the roof rafters, which must be able to carry the deck's load, but they should probably be at least 3½ in. tall at the narrow end so that they can be secured with joist hangers. A rim joist will be needed to cap the ends of the joists, but it should leave a gap at the bottom for drainage. A fascia that hangs down lower than the rim may be a good way to hide the rim and any exposed membrane or flashing. You can break

## A Typical Deck Over a Flat or Nearly Flat Roof

*The construction is similar to a regular deck, but the joists must bear on and align with the roof rafters rather than on a beam and posts.*

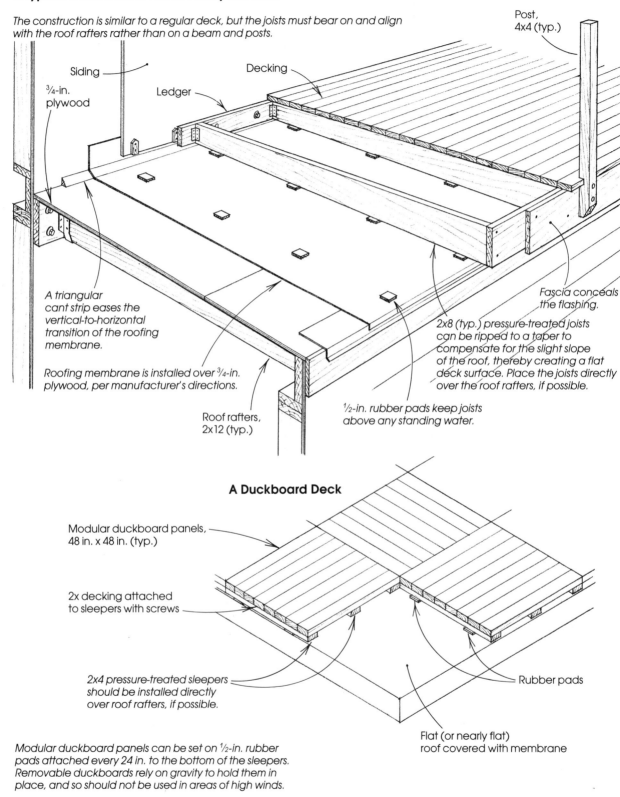

Post, 4x4 (typ.)

Siding

¾-in. plywood

Ledger

Decking

A triangular cant strip eases the vertical-to-horizontal transition of the roofing membrane.

Roofing membrane is installed over ¾-in. plywood, per manufacturer's directions.

Fascia conceals the flashing.

2x8 (typ.) pressure-treated joists can be ripped to a taper to compensate for the slight slope of the roof, thereby creating a flat deck surface. Place the joists directly over the roof rafters, if possible.

Roof rafters, 2x12 (typ.)

½-in. rubber pads keep joists above any standing water.

## A Duckboard Deck

Modular duckboard panels, 48 in. x 48 in. (typ.)

2x decking attached to sleepers with screws

2x4 pressure-treated sleepers should be installed directly over roof rafters, if possible.

Rubber pads

Flat (or nearly flat) roof covered with membrane

*Modular duckboard panels can be set on ½-in. rubber pads attached every 24 in. to the bottom of the sleepers. Removable duckboards rely on gravity to hold them in place, and so should not be used in areas of high winds.*

the deck into sections, which will make it easier to re-move to gain access to the roof in the future. This can be done by doubling some of the joists and having the decking break continuously between them and by us-ing easily removable screws on the joist hangers.

Posts on a roof deck with 2x8 or larger joists can be attached to the deck in the usual way (see Chapter 6) to either side of the rim joist. With smaller joists, or when using duckboards (described below), the posts will need to be affixed to the fascia of the rafter for a sturdy base. This means extra flashing is needed to keep the fascia and posts from being continually soaked by roof drainage. If it isn't aesthetically desir-able to have posts lapping the outside of either the deck or roof fascia, they can be moved back and be at-tached to the roof framing. This requires a special flashing that penetrates the membrane and is usually installed by the roofing contractor.

Another technique is to build a duckboard deck, which is a series of modular panels with integral pressure-treated 2x4 sleepers rather than joists. The duckboards can be constructed in the largest size you can lift and are usually laid right on the roof. The sleepers should align with the roof rafters, so choose a decking board size and pattern with this in mind. You can hide all the fasteners by building each panel up-side down and screwing through the sleepers into the decking. Be sure to allow a passageway for water around all the edges of each panel. Duckboards usually rely on gravity to hold them in place and shouldn't be used where high winds are the norm. Nonremovable duckboard sleepers can be held down by setting them in a bed of roofing tar on a bituminous roof.

To help protect against chafing on the main mem-brane, it's a good idea to add an extra strip of roofing material between any joists or sleepers and the roof membrane. This reinforcement should be glued down with a compatible glue. On bituminous roofs, the re-inforcement can be a strip of rolled roofing attached with roofing tar. Gravel on any of these roofs should be removed or sealed over as the weight of the sleepers or joists will eventually force it down and cause leaks.

Air circulation can be improved by attaching 3-in. by 3-in. by ½-in. pads of neoprene rubber (or whatever material is compatible with the roofing membrane) to the bottoms of the joists or sleepers, spaced every 2 ft. The pads keep the wood out of the water and help bridge over any irregularities in the roof's surface.

# Putting Down Decking

## Chapter 5

For me, laying the decking is a bit like putting the icing on a cake. It's generally easier work than the foundation and framing chores, and it provides some real visual satisfaction. Once the first few boards are down, the structure starts to look like a deck.

## Choosing the Decking

In Chapter 2, I discussed decking materials and fasteners at length, so you might want to review that information before getting started. Here I'll summarize some of the more important points and then focus on how to put the decking down.

The decking is a very visual part of a deck, and it gets a lot of direct contact from its users. These points should be considered when selecting decking boards. Boards intended for decking can usually be bought with the four edges slightly rounded, which helps reduce splitting along the edges and cuts down on potential splinters. Whenever possible, I like to use cedar or redwood for the decking—they're much more attractive than pressure-treated pine or fir. But cedar and redwood usually cost more (in some locations, a lot more) and they probably won't last quite as long as good, treated lumber.

**Widths**   Although lumber is readily available in nominal widths up to 12 in., it's unwise to use wide boards for decking. Boards expand and contract across the grain as they pass through the wet/dry cycle. The wider the board, the greater the movement. This movement results in larger cracks between adjacent boards during their driest stage and greater tension on

the fasteners, which will work themselves loose more quickly as a result. And if the fasteners don't give a little during these cycles of movement across the grain, wide boards will split and check. Finally, cupping looks worse on wide boards. Narrow boards will cup, too, but the effect will be much less noticeable when spread over several narrow boards.

The most common widths used for decking are 4 in. and 6 in. (nominal). The widest board I would use for decking would be a 2x8, and then only on a special design. I use 2x6s almost exclusively unless the clients have their heart set on some pattern requiring 2x4s. I feel that 2x6s avoid the problems associated with wider boards, while also offering several advantages over 2x4s. For one thing, 2x6s can be laid considerably faster than 2x4s. Each size requires the same amount of work and the same number of fasteners, but the 2x6 covers two more inches than the 2x4. Also, a defect in a 2x6 is less likely to cause failure—there's usually enough good wood surrounding the defect to provide adequate support. It's easier to locate a suitable nailing spot on a 2x6—you have more room to avoid knots. A knot on the end of a 2x4 may allow for only one nail, or it may have to be cut off. Choosing the optimum place to put the nail can also mean fewer split boards at the ends.

Two-by-fours are fussier in general, too. They tend to have more curves and warps over their total length, especially if they haven't been stored properly and were allowed to dry in a jumbled pile. They're often cut from smaller trees and are less stable than larger boards. Defects that you could live with in a 2x6 often need to be removed in a 2x4, and that takes time. Splits are more common in the ends of 2x4s.

When you're using pressure-treated pine or fir for decking, good-quality 5/4 boards are often a good choice. These boards are only 1 in. thick, but they're comparable in strength to 2x cedar and redwood.

**Lengths** Using longer lengths of decking boards means handling fewer pieces, and that saves time. Longer boards mean fewer butt joints, which means less potential rot and fewer split ends. Longer boards usually don't cost any more per linear foot until you get beyond 16 ft. It's worth some extra expense to buy long boards if butt joints can be eliminated from the deck. I try to use 14-ft. or 16-ft. lengths, but often I'm at the mercy of the lumberyard.

You should choose deck-board lengths with the goals of minimizing waste and labor without compromising the appearance of the deck. Deck boards must extend past the side joists to cover the fascia, if you're using one, plus another inch of overhang and then an additional inch or two to allow for trimming. This means that the total run of decking will be about four inches longer at each end than the width of the framing. If you're building a deck on a 19-ft. wide frame, you could use 10-ft. deck boards and not have much waste to cut off at the ends. This would require only one double joist in the center, but the butt joints would fall in a straight line, which is not an attractive option.

A more attractive choice might be to order 14-ft. boards and then lay each row with one 14-ft. board and one 7-ft. board made by cutting the 14-footers in half (as shown in the drawing below). This would require two sets of double joists and allow the joints to be staggered, which would result in slightly more waste. One row would be a 7-footer and a 14, the next a 14 and a 7.

The problem with this kind of planning, however, is that the lumberyard often has only a couple of lengths of decking boards in stock and you'll have to make the best of what's available. The shortest deck-board length I would use would be 4 ft., and I try to use nothing smaller than 6-ft. boards. Shorter pieces of decking look like patches, and they tend to self-destruct faster than long boards.

**How much to order?** Each linear foot of a 2x4 covers 0.29 sq. ft. (3.5 in. divided by 12). So if you're using 2x4s exclusively, divide the total square footage of the deck by 0.29. On a 12-ft. by 20-ft. deck (or 240 sq. ft.), you will need 828 linear ft. Likewise, a linear foot of 2x6 decking covers 0.46 sq. ft., which means that a 240-sq.-ft. deck would need 522 linear ft. of 2x6 decking. If you're using 14-ft. boards and staggering the joints as shown below, you would need to buy for a 252-sq.-ft. deck (12 ft by 21 ft.).

These figures don't take into account any waste, or the space between boards. In my experience, as long as I'm laying a simple straight pattern, the deduction for the space equalizes the waste; so I just leave both out of my calculations. If you're laying a fancy pattern or if the boards require a lot of trimming to remove splits, you should add an extra 5% to the order. I don't usually add any extra for waste; I prefer to wait until near the end of the job to determine the extra boards that I need. With more complicated decks, I often find that the plans change enough during construction to alter my original estimate anyway.

## Decking Patterns

The most common way to lay decking is parallel to the wall of the house—and usually it's the most attractive way. I think it's often a waste of time to try and spice up a deck with a fancy decking pattern. If there's a little extra fat in the budget, I prefer to spend it on nicer

**Planning Decking Needs**

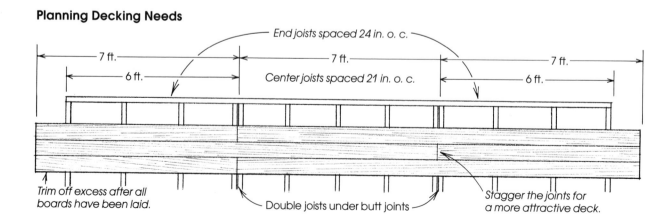

End joists spaced 24 in. o. c.

7 ft. — 7 ft. — 7 ft.

6 ft. — 6 ft.

Center joists spaced 21 in. o. c.

Trim off excess after all boards have been laid.

Double joists under butt joints

Stagger the joints for a more attractive deck.

## Decking Patterns

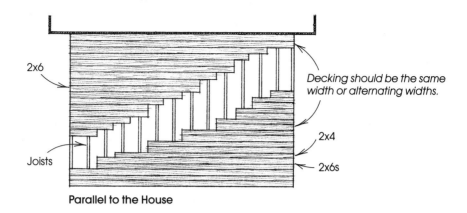

2x6

Joists

*Decking should be the same width or alternating widths.*

2x4

2x6s

**Parallel to the House**

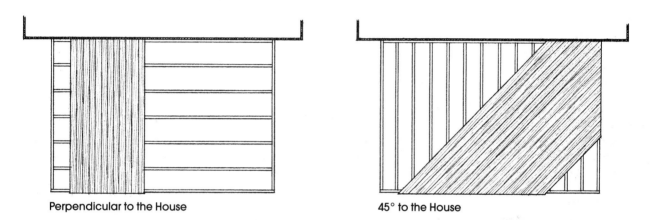

**Perpendicular to the House**

**45° to the House**

## Combination Patterns

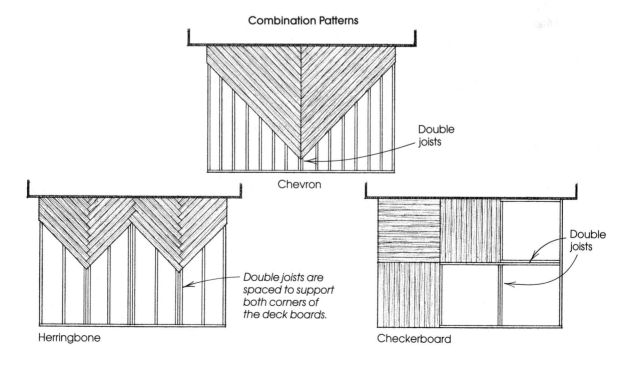

Double joists

**Chevron**

*Double joists are spaced to support both corners of the deck boards.*

**Herringbone**

Double joists

**Checkerboard**

Add some contrast to the decking by reversing the direction of the boards on a level change. This is also a safety feature because it calls attention to the step.

wood (cedar or redwood). Likewise, I like to devote my creative impulses to details that have more "curb appeal," such as railings, nice stairs, angles or level changes. A plain deck is rarely improved by an intricate decking pattern.

There are times when you may need to run the decking perpendicular to the house, perhaps because of some aesthetic desire of the client. This may be useful on a deck that will need to be shoveled regularly, for example. This orientation may also allow the use of shorter deck boards and perhaps even eliminate the need for butt joints. But this decision needs to be made when you start building the foundation, for the whole framing system needs to be rotated. (The framing for perpendicular decking is shown on p. 95.)

Another option is to run the decking boards 45° to the joists. This definitely takes a lot of work. Every butt joint must be cut at a 45° angle and is therefore longer and more likely not to fit the way you intended. There's more waste in this method because all the ends will be trimmed by at least 6 in. A diagonal pattern can usually be placed on a regular frame, though it's always wise to double joists under the butt joints to allow for better drainage. The joists may need to be spaced closer together because the deck boards need to span a greater distance from joist to joist. Diagonal decking is one of the slowest patterns to put down.

Some other pattern options might include a combination of parallel, perpendicular and diagonal orientations (see the drawings on p. 111). Any of these patterns could double the time needed to joist and lay the decking. Material costs will also be higher.

There are some simple tricks that you can play with the decking to add some contrast to a deck without adding much work or expense. Try alternating 4-in. boards with 6-in. boards, or two of one followed by one of the other. You can also add a level change, with the decking running in different directions on each level. Adjustments like this will keep the framing complications to a minimum while adding pizzazz and visually differentiating levels.

I've seen some decks with the deck boards installed on edge, which allows the boards to span farther than if they're flat, but these boards are difficult to attach to joists. This method requires a lot more lumber, and I don't find it particularly attractive.

## Trimming the Edges

Before you start laying the decking, you should give some thought to how the edges of the deck will look. There are several ways to deal with the exposed edges. One option is to cantilever the deck boards over the rim joist. The boards should hang over about 1 in., although a little more than this looks better than a little less. This system is quick, as the boards can be left long while they're being installed and then cut to length all at once. If the rim wasn't installed perfectly straight, cantilevered decking will mask the problem. But some folks don't like the look of all that exposed end grain on a deck. Crosscuts are often fuzzy, and end grain tends to look rougher over time. These cut ends also require frequent applications of water repellent.

One way to add a distinctive look to the deck while covering the ends is to add a "picture-frame" edge band to the decking. The band can be the same width or wider than the deck boards and is laid flat, overhanging the rim. The corners should be mitered, but try to avoid a tight fit at the cut ends so that they can dry out. This treatment requires an extra joist in the framing and some extra blocking.

Another alternative is to cover the framing and the decking with a wide 5/4 or 2x fascia. You could use 1x stock, but the edge will be more fragile. With this method, all the deck boards should be installed first and trimmed flush with the edge of the rim joists below. Then install the fascia board with its top edge flush with the finished surface of the decking and the ends mitered.

The fascia can also be installed under the deck boards, covering only the framing. I prefer this method because I like the traditional look of overhanging deck boards. When all of the edges are covered with a solid fascia, the deck looks too streamlined and modernistic for my taste.

A variation on either approach is to break a wide fascia into two overlapping layers. The offset adds some visual interest by creating a shadow line. With either method, the fascia is often nailed directly to the face of the joists. This makes for easy installation, but also makes it difficult for the faces of the boards to dry, and it's often the first place that I see rot on a deck. A better approach is to attach the fascia with pressure-treated plywood spacers to aid drainage (for more information on spacers, see p. 84).

Spacers should be placed at each railing post location. Since the post locations won't be known yet, it's wise to wait and put the fascia on after the railing has been laid out.

## Which Side Up?

An old adage tells you to lay decking with the "bark side" up, which means that the growth rings are oriented in a dome-like rather than cup-like fashion. This advice is often sound because of the way wood behaves as its moisture content varies. Briefly, moisture causes wood to move more parallel to the growth rings than it does perpendicular to them. So, when a flat and dry board gets wet it will expand more on the bark side than on the pith (center) side. Thus, any cupping that happens would result in water draining off the board. If the same board is installed upside down, the cupping would result in retaining the water on the board, which would promote rot.

**Picture-Frame Edging**

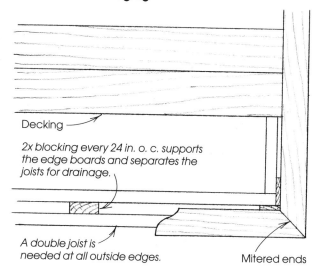

Decking

2x blocking every 24 in. o. c. supports the edge boards and separates the joists for drainage.

A double joist is needed at all outside edges.

Mitered ends

**Installing the Fascia**

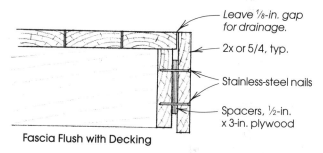

Leave ⅛-in. gap for drainage.

2x or 5/4, typ.

Stainless-steel nails

Spacers, ½-in. x 3-in. plywood

Fascia Flush with Decking

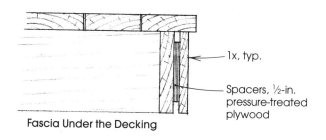

1x, typ.

Spacers, ½-in. pressure-treated plywood

Fascia Under the Decking

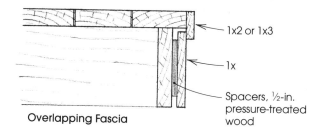

1x2 or 1x3

1x

Spacers, ½-in. pressure-treated wood

Overlapping Fascia

## Bark Side Up?

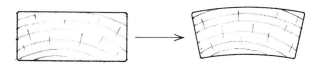

*For wood that's prone to cupping, it's wise to install the decking bark side up. That way, if it does cup, the water will run off.*

However, if the wood is milled flat and installed bark side up while still green, it's possible that it will shrink parallel to the growth rings and end up cupped to retain the water, exactly the opposite of what the adage would have you believe. What all this means is that you should dry your wood before you install it if you want to lay it bark side up.

Installing the decking bark side up is a good suggestion if both faces of the board are of the same quality. In my experience, this is often not the case. One side of the board will have some kind of defect that convinces me to put that side down. This is especially true with cedar. I follow a different adage and put the best side up.

Neither cedar nor redwood has an excessive tendency to cup, which makes it possible to ignore the "bark-side-up" rule without unfortunate consequences. With pressure-treated pine or hem-fir, the situation is different. These softwoods do like to cup, so you would be well advised to use dry wood and put the bark side up. Ignore this advice only when the results are too ugly. Fastening the boards well and maintaining the wood with a water-repellent finish will help prevent many cupping problems. You can always replace a board or two later if they rot prematurely because of excessive cupping. The narrower the board, the less severe the cupping problems, so stick with 2x4s and 2x6s.

## Installing the Decking

The best favor you can do yourself when laying decking is to get the first board down as straight as possible. I snap a chalkline on top of the joists for the first board. This helps to ensure that the board will go on straight regardless of irregularities in the house wall. If you're laying the boards parallel to the house, measure out from the wall the width of the first piece of decking plus ¼ in. (to allow for drainage between the board and the siding or flashing) and make a mark on the top of the two end joists. Then snap a chalkline between the two points. Set the first deck board in place and nail it down, working from one end to the other, keeping it flush with the chalkline. You'll have better luck by starting with the straightest board you can find.

If you're installing the boards perpendicular to the house, the chalkline should allow for the overhang. If you're laying the decking diagonally, don't start at a corner with a small piece. Instead, begin with a board that's at least 4 ft. long. This will make it easier to ensure that the first board is at 45° to the joists.

All butt joints should be aligned and gapped carefully, but let the ends of the deck boards run wild over the edge for now. When you've finished fastening down the decking, you can snap a chalkline and cut the ends all at once. If the cut runs along the top of a joist, be sure to set the saw blade to minimize cutting the joist. If the cut ends up next to a wall, you may have to complete it with a handsaw. To avoid this, I try to remember to install the first board at its exact finished length.

It's a good idea to think about the last row of decking before you reach it. You don't want to end up cutting a skinny board just to fit the edge. If I'm laying decking with a 1-in. overhang, I always try to keep at least 2½ in. to 3 in. bearing on the joists. So, a few rows before reaching the end, do some measuring to see how the boards will end up. You may be able to adjust the gap a little on the last few rows, thus avoiding the need to rip any decking boards. If that won't work, you might consider ripping small amounts off the last couple of rows, which won't be too noticeable. My favorite solution is to use a single, wider board that is ripped to the width I need.

First-time deck builders often think they should divide the surface up into equal spaces and then adjust the gaps between the boards so that they end up with full-width boards on the whole deck. But it's not a perfect world, and slight variations in board width will almost certainly wreak havoc with all of your careful calculations. Instead, just lay the boards and worry about how to the deal with the last few rows when you get there. It always works out and no one ever notices that the last board or two is slightly narrower or wider than the rest.

## Getting the First Board Straight

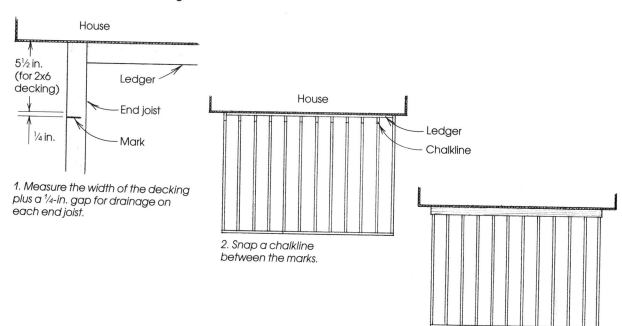

**5½ in. (for 2x6 decking)**

**House**

**Ledger**

**End joint**

**¼ in.**

**Mark**

1. Measure the width of the decking plus a ¼-in. gap for drainage on each end joist.

**House**

**Ledger**

**Chalkline**

2. Snap a chalkline between the marks.

3. Lay the first row of decking flush with the chalkline.

When installing the decking, let the joist ends run wild. When you've finished nailing, snap a chalkline along the edge and cut all the boards at once. Note that the decking overhangs the joist by about 1 in.

## Fastening the Decking

There are several ways to fasten the deck boards to the joists. They can be nailed with galvanized or stainless steel nails, preferably with a ring shank. Or they can be screwed, again using galvanized, stainless-steel or specially coated screws (see Chapter 2 for more information on fasteners). You can also use deck-board clips or ties, which I'll discuss below.

The quickest and most common fastening method is nailing, either with a hammer or a pneumatic nailer. For 2x decking, you should use 16d nails; for 5/4 decking, use 10d nails. When nailing, take extra care at the board ends—a nail can split the board. The surest way to avoid this is to predrill a hole somewhat smaller than the diameter of the nail. By drilling at an angle, you can keep the head of the nail farther from the end of the board. A somewhat less predictable technique is to blunt the point of the nail with a hammer before driving it into the deck board.

The usual pattern is two nails per decking board at each joist. Keep them about ⅝ in. from the edge of the board for 3½-in. wide boards and a little farther on 5½-in. boards.

The California Redwood Association recommends fastening wet redwood with only one nail at each joist crossing, staggering the pattern from one side of the board to the other. This approach supposedly allows the board to dry and shrink in width with less restraint, thus reducing the chances of splitting. I've never run into this problem with wet wood, but the logic seems reasonable and might prove useful with any wood. If you follow this advice, you'll want to go back in a year or so and finish the nailing. Most pressure-treated softwoods, however, are best nailed completely at installation time to minimize cupping and warping.

Don't use a serrated-faced hammer on the decking. It will make the deck look like a giant waffle. With either a hammer or pneumatic nailer, take special care not to damage the decking or drive the nails too deep. But don't go overboard—minor dings in the wood will swell back up in time.

Nails should be set just a little below the surface by driving them until they are almost flush with the decking and then finishing with a nail set. Because the wood will likely shrink after installation, you'll want to go back and reset them in a year. This should suffice for ring-shank nails, but smooth-shank nails may need to be reset every year. The depth of pneumatically driven nails can be adjusted somewhat by controlling the air pressure on the compressor, but they'll almost certainly be driven deep enough that they won't require future attention.

**When nailing at the ends, take special care not to split the board. A safe way to nail the ends of boards is to predrill the hole, at any angle, before nailing.**

Installing screws is slower. If the screws have a section of unthreaded shank just below the head, then you won't have to worry about predrilling holes except at the ends of boards. The unthreaded part of the shank allows the deck board to be pulled tightly to the joist. If your screws are threaded all the way to the top, try a few of them before deciding whether or not predrilling is necessary. Predrill only the deck board, not the joist. The diameter of the bit for predrilling should be about two-thirds the outside thread diameter of the screw. It's much faster to use separate drills for drilling the holes and for driving the screws. Drive the screws until their tops are just below the decking. They can be driven deeper later if the wood dries and leaves them proud.

If you want to use screws and speed up the process, there are a couple of special tools available (see the Resource Guide on p. 150). The Quik Drive automatic screw-driver system attaches to a standard electric drill and drives precoiled screws that load into a self-feeding magazine. While the Quik Drive system is useful for other applications, such as installing drywall, you're limited to using the manufacturer's screw coils. This is not a tool for the one-time deck builder. The Dec-U-Drive is another drill-powered tool, but it operates with loose screws. This tool can be used while you stand up, but I would consider this to be a disadvantage since you really need to be on your knees, prying on the boards to get them straight, while fastening.

I've seen adhesives recommended in some books on deck building, but I think they're a waste of time. I wouldn't trust a construction adhesive to hold up outside for more than a year or two. Besides, even if you use an adhesive, you'll still have to use a lot of fasteners to hold the boards in place while the adhesive sets.

A recent innovation for holding down the decking is the deck clip or deck tie. These are available from several different manufacturers in slightly different configurations (see the Resource Guide on p. 150). All of them are fastened to the underside of the deck board and therefore eliminate nailing through the top of the decking. They automatically space the deck boards, and at least two models hold the decking off of the joist, thereby eliminating direct wood-to-wood contact. If you've spent a lot of money for some clear lumber to use as decking and you'd rather not mess up the surface with a lot of nails, then deck clips are the way to go.

There is a price to be paid for this improvement, however. Deck clips slow installation of the decking appreciably, and they cost a lot more than nails. Because the clips automatically space the boards, they probably shouldn't be used on wet lumber since the

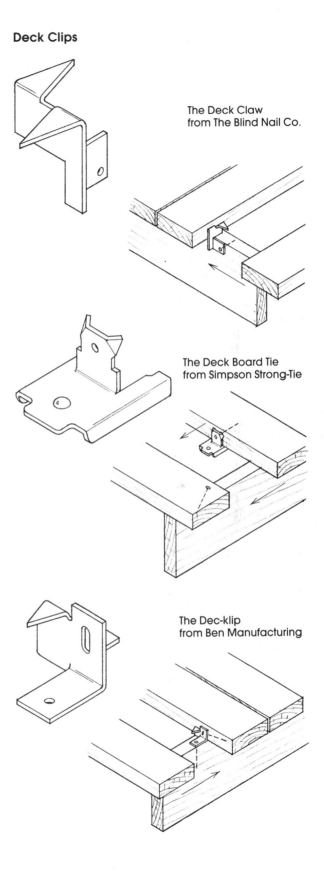

**Deck Clips**

The Deck Claw
from The Blind Nail Co.

The Deck Board Tie
from Simpson Strong-Tie

The Dec-klip
from Ben Manufacturing

gaps between the boards will be too wide when the wood dries. Also, shrinking lumber will loosen the grip of the clips and make them less effective. Another potential complication with deck clips is with butt joints, where the clips would have to be installed right near the end of the boards, which would create splitting problems. The problem can be overcome by using a double joist or a wider joist to keep the clip away from the end.

Like the others, the Deck Claw made by The Blind Nail Company is galvanized, but the manufacturing process leaves the points that go into the boards unplated and highly likely to rust away long before pressure-treated wood rots. The Deck Board Tie made by Simpson Strong-Tie is lightweight and has very small prongs that won't grip well if the board shrinks. The Dec-klip is the sturdiest, holds the boards off the joists and is plated after manufacturing, which ensures better rust protection. It's biggest disadvantage is that it grips the joist about ½ in. from the bottom of the deck board, which could cause problems with lower-quality 2x stock.

## Spacing the Boards

Some builders use a carpenter's pencil to space deck boards. With wet boards, however, this results in some huge gaps once the wood dries out—as much as ½ in. I try to shoot for ⅛-in. gaps after the boards have dried.

A 16d nail is a readily available spacer that can be used for reasonably dry deck boards.

This is more than adequate for water to drain through and should allow for air circulation as well. I also think that ⅛-in. gaps look best.

Gapping deck boards is far from an exact science. There's a lot of variation in the ways different batches of wood dry. I've developed something of a sixth sense for things like this. If the decking hasn't been kiln dried and is heavy and wet, I usually lay the boards tightly against each other with no gap at all. They'll shrink, even in damp climates, and leave an adequate gap, although this may take a year. With kiln-dried boards or boards that have been sticker-stacked long enough to dry, I usually use a 16d nail to space them. I drive the nail through a piece of plywood, which makes it easier to pull it out when the boards have been nailed. I wouldn't space the boards any closer than this, as they'll swell when they get wet. If you like gadgets, you might enjoy using a Deck Mate (see the Resource Guide on p. 150), which will help you consistently space the boards ⅛ in. or 3⁄16 in. apart. When the boards are somewhere between wet and dry, it can be hard to guess how to gap them. The folks at the lumberyard may be able to help.

## Problem Boards

No matter how much you paid for your decking lumber, some of the boards are going to be crooked. Cedar and redwood are seldom cupped or twisted, but they may have a curve or bow from one end to the other. Pressure-treated pine or fir is likely to have some of these problems, especially if it hasn't been stored properly.

Boards that are badly cupped should be discarded— it's virtually impossible to flatten them out. A bowed board can be stubborn, but most of them can be installed with a few simple techniques. The most common method is to start at one end and nail into as many joists as you can until the bow gets too pronounced to keep in line by hand. You should be able to nail into two or three joists without much sweat.

When the bow gets to be too much to handle by hand, I usually grab my 1½-in. Stanley framing chisel. I drive the chisel into the joist angled away from the deck board. Then the chisel can serve as a lever to force the board into line. If you're using a hammer to fasten the decking, you'll need to start the nail first. You can also use a flat pry bar, such as the Stanley Wonder Bar, for a lever.

If I can't pry the board into position, I'll generally do one of two things. If I don't mind adding an additional butt joint, the quickest solution is simply to crosscut the board in half, which divides one long

**For boards that need a little persuasion, I use a chisel to pry them into line. It'll be easier if you start the nail in the board first.**

**The board edges will look nicer and be less likely to splinter if they're rounded over with a router.**

board into two more manageable pieces. If I don't want this additional joint, then I'll try to wedge the board into position. To do this, first nail a 2x block across the tops of two joists, parallel with the deck boards and out near the end of the bowed board rather than in the middle. Then cut a wedge out of some scrap material with only a slight taper, say 1 in. to 2 in. in 12 in. Drive the wedge in a little at a time as needed to complete your nailing.

Pipe clamps can be useful if they're long enough and if there is something to hook them on to. In my experience, these are two big "ifs." For those of you who build a lot of decks and run into a lot of crooked decking, you might appreciate the Lumber Jack (see the Resource Guide on p. 150), which is a lever-action, ratcheting tool made especially for convincing crooked

boards to straighten out. As a rule, cedar and redwood are easier to straighten than pressure-treated lumber.

## Finishing Touches

After all of the decking has been installed, I like to round over all of the edges with a router to eliminate some of the fuzzy edges and give a more handcrafted look to the deck. A ⅜-in. roundover bit is usually sufficient. A 45°-chamfer bit works just as well.

One last bit of information: The siding on many beautiful houses has been ruined by the addition of a deck. Why? Because rain dripping off the roof now hits the deck and splashes onto the siding. The solution? Install gutters.

# Railings

## Chapter 6

In my opinion, nothing more influences the overall appearance and appeal of a deck than the railing. Railings are highly visible features that prevent people from falling off the deck—they also allow the builder to be creative by turning an otherwise basic deck into a structure to be proud of.

Railings can be constructed in a wide variety of shapes and sizes using many different materials. Wood is far and away the most popular material, but various metal tubes and cables, plastic tubes and sheets, and ceramic tile can be good choices for railings, usually in combination with wood.

The starting point for any railing design must be safety, however, not aesthetics. Deck railings must conform to certain minimum standards, and building codes are very specific about these standards. Railings also need to be strong, as they are subject to extreme lateral forces. Vertical members such as posts often must act as levers to concentrate significant loads directly on the bolts holding the railing together.

Because they're so visible and so frequently touched, railings need to be regularly maintained. And since they're usually made up of many smaller pieces, the maintenance needs more time and effort than other parts of the deck. I think wood railings are the most attractive, but I've found no real shortcuts to keeping them looking that way. I recommend applying finish to the railing, or at least to the top rail, every year.

Before designing the railing, you need to consider what kind of view to expect from the deck. Some clients may want a deck that's concealed from its surroundings, such as nearby neighbors. In this case, the railing needs to be very full. On the other hand, a deck railing may need to maximize a beautiful view, in which case you would want as unobtrusive a railing as possible. Here, thin cable or sheets of clear acrylic might be good choices.

**Railing anatomy** Railings are made of several parts. Most traditional wood railings have posts for support, with horizontal rails and numerous smaller vertical balusters. Wooden posts are either lapped over the outside of the rim joist or attached between joists on the back side of the rim. The top rail may be composed of several parts, in which case the uppermost piece is called the cap rail. Railings often have a bottom horizontal rail. If you want to move away from traditional railings, the large intermediate posts can be eliminated and the railing can rely on balusters lapping the rim for support.

Another design option is for the posts to be the only vertical members, with the intermediate spaces filled with horizontal rails. Or railings can be short, solid "pony" walls covered with siding to match the house. And almost any part of the railing can be made out of some material other than wood. Posts and rails can be made of metal pipe, and balusters can be steel rod, tube or cable. Individual balusters can be replaced entirely by large sheets of wire mesh, clear plastic or lattice. Although we continue to use traditional terms for the parts of a railing, in many systems the materials and installation may be quite different.

## Code Requirements

Codes for railings vary from place to place, and they're strictly enforced. Check with your local building inspectors before you get started. Most railing requirements address two issues: the height of the rail and the gaps between the railing components.

Railings are usually required whenever the deck is more than 30 in. above the ground. To me, this is a bit too lenient. This regulation takes care of observant adults, but it doesn't protect children, especially small ones. If the deck is to be used by kids (and how many aren't?), I recommend installing a railing anywhere that one could fall more than the thickness of the deck. In other words, unless the deck sits on the ground, there should be some sort of restraint at the edges. For smaller drop-offs that aren't regulated by code, you might use a shorter-height railing or a series of planters and benches. Sometimes it's adequate to announce the drop-off with some visual contrast, such as a painted stripe, a raised bump or small lighting fixtures.

I've never seen a regulation that allows a single-family residential railing to be less than 36 in. high. Commercial and multifamily units are likely to need 42 in., and many areas are applying the higher requirement to all structures. It's not hard to imagine someone unfamiliar with the terrain flipping over the top of a 36-in. high railing. The higher railing definitely feels safer, and it adds very little to the cost. I recommend a higher railing when it doesn't interfere with the view, and especially when the deck is more than 8 ft. off the ground.

For railing components, the Uniform Building Code has adopted a maximum spacing of 4 in. between elements. That is, nowhere in the railing system can a 4-in. diameter ball (which is meant to represent a child's head) slip through. The requirement may be 6 in. in some areas, but I suggest that you follow the 4-in. rule even if it isn't required yet in your area. If you don't do it now, it could mean additional work if the house is resold. Also, some codes require that the space between the bottom rail and the deck not exceed 2 in.

## Wood Railings

I build railings almost exclusively out of wood. When building a deck, the crew is set up for carpentry and woodworking, so it's easiest to continue with that material. Wood railings have evolved a lot over the years from traditional porch-like systems to more complex styles and delightful combinations of traditional styling and creative variations.

I use the same wood for the railing that I use for the decking. For me, this is usually cedar or redwood, but it could just as well be pressure-treated pine. Most of what I say about railings will apply no matter what type of wood you choose.

When I begin laying the decking, I "high-grade" the lumber pile. This means that I set aside the choicest boards to use on the railing. The clearest, straightest and most vertically grained pieces will be best able to stand the tests of time and weather.

A common dimension needed for railing balusters is a nominal 2x2, which measures 1½ in. by 1½ in. These are usually made from 2x stock and therefore only have to be cut once to get a square piece. It's possible to buy 2x2 baluster stock from the lumberyard, but I try to avoid it. The 2x2s stocked by lumberyards have

### Code Requirements for Railings

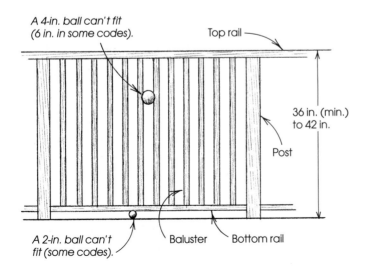

A 4-in. ball can't fit (6 in. in some codes).

Top rail

36 in. (min.) to 42 in.

Post

A 2-in. ball can't fit (some codes).

Baluster

Bottom rail

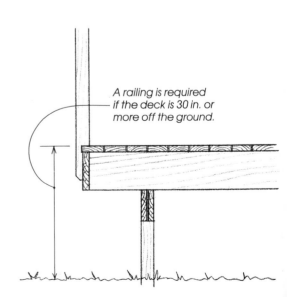

A railing is required if the deck is 30 in. or more off the ground.

## Simple Railings Using a Continuous Top Rail

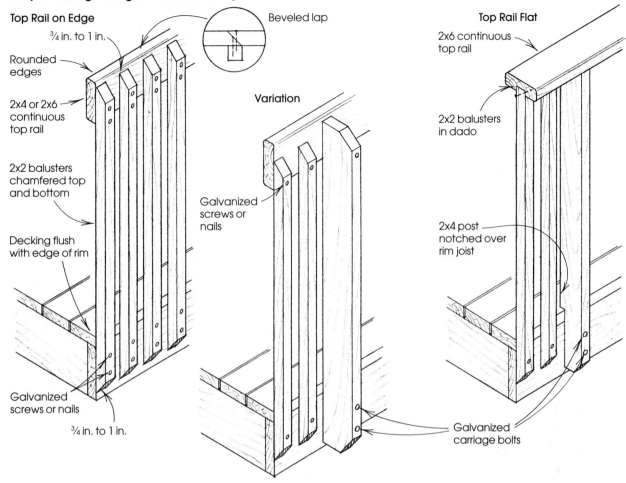

**Top Rail on Edge**

¾ in. to 1 in.

Rounded edges

2x4 or 2x6 continuous top rail

2x2 balusters chamfered top and bottom

Decking flush with edge of rim

Galvanized screws or nails

¾ in. to 1 in.

Beveled lap

**Variation**

Galvanized screws or nails

Galvanized carriage bolts

**Top Rail Flat**

2x6 continuous top rail

2x2 balusters in dado

2x4 post notched over rim joist

Galvanized carriage bolts

often been allowed to dry helter skelter, and they tend to be terribly warped and twisted. They also have often been salvaged from larger boards containing defects, and therefore can be somewhat defective themselves. And even if these problems don't exist, the 2x2s usually won't have the same visual characteristics (such as color) as the rest of the railing material or the deck.

For the best results with cedar and redwood, I suggest that you saw your own 2x2s out of the same lumber that was used for the decking and railing posts. If you're using pressure-treated lumber and you can't find good-quality 2x2s, you'll have to apply extra preservative to any cut edges. When setting aside boards for later use it's easy to underestimate how much you'll need for balusters—they may be small, but they're numerous. If you design your wood railing to use standard sizes (4x4s for posts, 2x4s and 2x6s for the railings), you'll save yourself a lot of work.

When sawn with a sharp blade, dry softwoods require little extra work to be put to use. At most, you might want to touch them up with a belt sander. When this is necessary, I put my belt sander on its back, secured with a spring clamp, and then hold the wood to the sander, rather than the other way around. This way, I get a much better feel for how much wood I'm taking off.

**Railing styles** The simplest railings are comprised of a series of balusters attached at the bottom to the rim joist and at the top to a 2x4 or 2x6 railing installed on edge (see the drawings above). The top rail can also be placed flat. As long as the balusters are installed firmly, this system doesn't require intermediate posts for support. The top rail should be made with the longest pieces available and run all around the perimeter of the deck. When the top rail is on edge, the joints be-

## Railings With Intermediate Posts

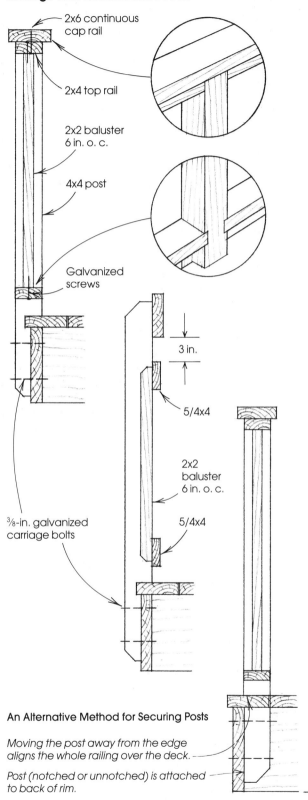

2x6 continuous cap rail

2x4 top rail

2x2 baluster 6 in. o. c.

4x4 post

Galvanized screws

⅜-in. galvanized carriage bolts

3 in.

5/4x4

2x2 baluster 6 in. o. c.

5/4x4

### An Alternative Method for Securing Posts

*Moving the post away from the edge aligns the whole railing over the deck.*

*Post (notched or unnotched) is attached to back of rim.*

tween boards in the railing will be less conspicuous if they're beveled (that is, lapped with 45° cuts on both boards). The railing joints should always fall over a baluster so that they can be fastened securely.

To break up the monotony of all the balusters, it's common to install a vertical 2x4 or 2x6 on edge every 6 ft. or so, as shown in the middle drawing on p. 121. A 2x4 can be attached directly with lag bolts. A 2x6 should first be notched 1½ in. to keep it from sticking out too far. These boards can also be nailed or screwed on from the back side through the rail or rim. This style of railing works just fine for those with limited time and money.

I used to build a lot of railings like this, but I felt something lacking from the design. I began to look at railings on porches of older homes and I noticed that they rarely had balusters that reached down to the deck. Instead, they used intermediate posts for structural support and had the balusters attached to a bottom railing that was up off the floor. Since deck posts couldn't extend to a roof, as they did on porches, they had to be firmly affixed at their base. This type of post system is better looking, but it takes quite a bit more time to construct. This type of railing also makes the deck much easier to sweep clean of debris and snow.

The rail style that I use most frequently mimics this design. Balusters are connected to the top and bottom rails and not to the rim joist. This allows the decking to overhang the rim joist a bit, as discussed in Chapter 5. Each component (posts, rails and balusters) can differ in size and location from deck to deck, or even on a single deck, allowing a myriad of choices. Posts may be notched and lapped over either side of the rim joist, or sometimes not notched at all.

**Railing posts**   If the posts on this railing system are simply attached to the outside of the rim, it won't have any decking underneath it. This makes it look like it's floating away (which is not a reassuring feeling). The problem can be minimized by moving the railing in over the deck a little farther, which can be done by notching the bottom of the posts. Notching the posts also reduces the bulky look of large posts attached to the rim. For a 4x4, the most common size for posts, I cut a notch no deeper than 1½ in., leaving at least 2 in. to attach to the rim. If the decking overhangs the rim, as it does on most of my decks, the decking will have to be notched to fit the post.

Some designs need a less busy look, which can be achieved by moving the posts in farther on the deck. This positions the railing entirely over the deck. The post is attached to the back side of the rim, and it can be notched or not, depending on how far in you want it to be. The post passes through a hole in the decking, which adds a solid look to the deck.

I usually use 4x4s for posts. A 6x6 can be interspersed every so often to add diversity, but this needs a bit more planning. Choose your material carefully, as the posts will be very visible. I try to select my own post stock at the lumberyard rather than having it delivered sight unseen, and I often have to search several yards to find decent stock.

On some designs, the posts will be covered with a layer of 1x wood as part of a railing that incorporates decorative moldings and caps to complement the style of the building. This can result in a beautiful (but labor-intensive) railing. Turned posts and balusters are also available in different woods, and may look good on the right style of house.

You can choose between a continuous-top cap rail or a rail that butts into the posts. Starting and stopping a rail at each post means a lot more work. One way to add diversity without quite so much work is to have only a few of the posts go above the rail—these posts can have a decorative top to show off some nice workmanship, or could be tall enough to hold lighting fixtures. If you want a more streamlined and unified look, a continuous-top cap with equally sized posts all around might be a better choice.

The spacing of posts varies from deck to deck, depending upon the strength of the intermediate rail and baluster design. Posts must be close enough so that the railing will be strong and solidly attached to the deck. I usually space posts 5 ft. to 7 ft. apart, although they certainly could be closer together for aesthetic purposes.

If the railing is going to start and stop at each post, you'll need closer spacing than on a continuous-railing system. You may find it useful to build a small prototype section of railing to help you decide.

There are many ways to cope with railing corners. You can put a single post in the corner to hold the rails from both directions. This works well when the posts are inside the rim joists, but notching and bolting the bottoms of these corner posts to lap over the outside of the rim is more difficult and requires longer bolts and corner blocks inside the framing. I prefer to use two posts, each set back from the corner, that can be installed like all the others. Just remember that the space between these posts must meet code requirements. You could also place a short section of rail between the posts with a baluster or two. With this two-post corner, the continuous top cap rail is usually cantilevered over the top of the posts from each direction and mitered in the corner.

Whichever style of post fits your design, you'll need some blocking between the joists. If the posts are bolted over the outside of the rim, the block should be against the back side of the rim with a plywood spacer between them (as usual). If the post is going on the inside of the rim, it's often best to locate the blocking so that the post fits between it and the rim. In either case, the blocking is attached by nailing through the sides of both joists into the end of the block. If you don't have access to the underside of the deck, you'll need to install this blocking before completing the decking. Posts along the sides of the deck will normally be attached to a joist that is parallel to the other

## Turning Corners on a Railing

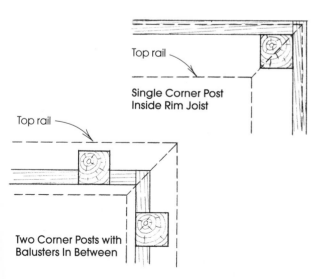

Top rail

Single Corner Post
Inside Rim Joist

Top rail

Two Corner Posts with
Balusters In Between

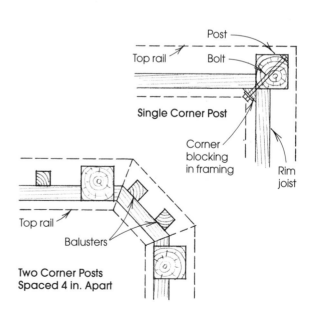

Post

Top rail    Bolt

Single Corner Post

Corner
blocking
in framing

Rim
joist

Top rail

Balusters

Two Corner Posts
Spaced 4 in. Apart

## Bolting Posts to the Rim Joist or End Joist

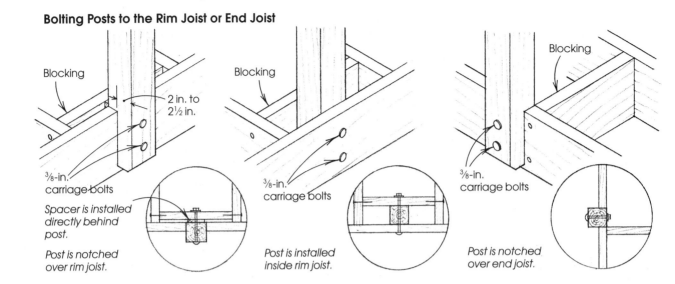

Blocking

2 in. to 2½ in.

Blocking

Blocking

³⁄₈-in. carriage bolts

*Spacer is installed directly behind post.*

*Post is notched over rim joist.*

³⁄₈-in. carriage bolts

*Post is installed inside rim joist.*

³⁄₈-in. carriage bolts

*Post is notched over end joist.*

joists. In this case, blocking should be installed between the joists, directly behind the post, as shown on p. 123.

Posts are usually attached with long carriage bolts though the rim, post and blocking. If access to the underside of the deck is limited, you may find it a good idea to leave a couple of pieces of decking loose for access until after the posts are installed. This decking could be installed with screws so that you could remove it to retighten the bolts later.

If the posts are all the same length, it's best to cut them all at once. Post length is determined by adding the height of the post above the deck surface to the amount below the deck surface. The height above will depend on the design, but remember that the top rail must be at least 36 in. above the deck to meet code. Generally, the bottom of the post should be level with the bottom of the rim.

A power miter box (or chop saw) with a stop on the fence will give the cleanest and squarest cuts. A table saw with a miter gauge will also work well, but it may require two passes to cut through a 4x4. If the posts run continuously from the foundation up to the railing, they'll need to be cut to length with a circular saw. Measure and mark the height on the end posts and then snap a chalkline between them to mark the intermediate posts.

If the post laps over the outside of the rim, then its bottom will be visible. In this case I like to bevel the bottom to make it more attractive. You could bevel just one or all three exposed corners. Just scribe a line using a square and cut the bevel on a power miter box or a table saw. Don't use a circular saw—it's dangerous

since the base doesn't have full support and you may be tempted to keep the guard up, too.

After the posts have been cut to length and the bottoms have been beveled, they need to be notched to fit the rim. Orient the notches so that the best side of the post is most visible. Notching the bottom of the posts is a somewhat tedious affair, but it needs to be done accurately. First, you need to lay out the notches. Rather than do a lot of measuring and marking, I like to use a framing square with a 1½-in. blade or a small try square with a 1-in. blade (either one of these dimensions is usually sufficient). Layout is then simply a matter of laying the square along the edge of the post and scribing a cutline. The only dimension that needs measuring is the distance from the bottom of the post to the top of the notch.

I usually use two circular saws to cut the notches—the first saw is set for a full-depth cut on both sides of the post, and the second saw is set to make the crosscut to the depth of the notch. The cut-out section can then be removed with a rap of a hammer. The notch will probably need to be cleaned up with a chisel.

If the posts are going to be bolted to the outside of the rim, this is a good time to lay out the bolt holes. I use two bolts per post and space them so that the upper one is about 1 in. below the top of the joist and the bottom bolt is as low as possible with the head of the bolt above the bevel on the post. If you have a lot of posts, you can speed the layout process along by using the simple jig shown in the drawing on the facing page.

The number of posts needed on a deck can be determined by laying out the post locations. First, choose

the spot for the corner posts and then measure the total space left over. Then subtract the combined width
of all the posts in that run. Divide the remaining space
by the number of spaces between posts to get the size
of each space. Now measure over from the corner post
that distance to begin the layout. Lay out all the post
locations on the deck before doing any cutting to
make sure your math was right.

If the posts are being mounted on the outside of the
rim and you have overhanging decking, you'll need to
cut notches in the decking with a jig saw to accommodate the posts. These should be the width of the
post and deep enough to be flush with the outside of
the rim. Posts that are going inside the rim will need a
square hole in the decking, which can also be cut with
a jig saw. If the posts are going inside the rim and
you're planning to cover the framing with a fascia, you
might want to put the posts on before the fascia. The
bolts will be concealed, but be sure to use carriage
bolts, since the heads won't be accessible for future
tightening.

Now place a post in position. If you don't have a
helper to hold the post, you can nail it temporarily.
Use a level to make sure the post is plumb, then drill
the upper hole through the post, rim joist and blocking. Use a spade or auger bit the same size as the bolt
used (I usually use $\frac{7}{16}$-in. galvanized carriage bolts to
secure posts to the framing, and I would never use less
than a $\frac{3}{8}$-in. bolt). Next, install the upper bolt with
nuts and washers and hand tighten. Check again that
the post is plumb, and then drill the bottom hole and
insert the lower bolt. Use a similar sequence for posts

The bottoms of exposed posts look much better with
beveled edges, which are best cut on a power miter saw
or a table saw.

The bottoms of railing posts should be notched to provide a better connection at the rim. Cut along the layout
lines with a circular saw, and then clean up the notch
with a chisel.

## Bolt-Hole Layout Jig

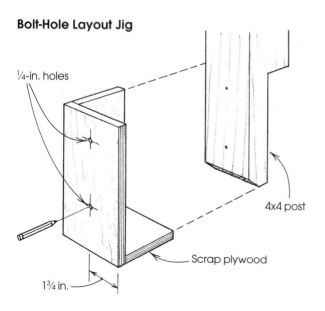

¼-in. holes

4x4 post

Scrap plywood

1¾ in.

For a tight fit and neat appearance, the decking over-hang must be cut out to accommodate posts that are go-ing on the outside of the rim.

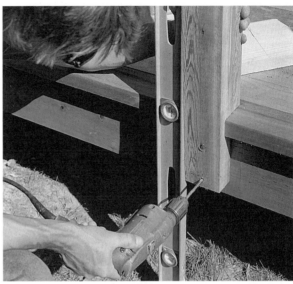

Attach the posts by first inserting the top bolt. Then, with the post plumb, drill the bottom hole through the post, rim and blocking. If the post isn't plumb front to back, trim out the notch or use shims.

that are inside the framing, but the layout will have to be on the outside of the rim. Always drill through the rim, post and blocking at the same time.

After tightening the bolts, check the inside and out-side of the posts for plumb. If the post isn't plumb, you can remove it and plane or chisel the notch to create a flatter connection, or you may need to shim it out with shims made with the same material as the posts.

## Between the Posts

One of my favorite ways to fill in the space between intermediate posts is with a balustrade made of upper and lower 2x4 or 2x6 rails and 2x2 balusters (shown on p. 122). The rails can be installed on edge with the balusters lapping their wide face or they can be in-stalled flat, with the balusters butting into the them. The rail can be continuous or it can start and stop be-tween posts. For best results with a single flat upper rail, cut a continuous groove the width of the balus-ters on the underside to prevent the balusters from turning in place. You can do this with a router or dado blade on a table saw. The balusters can then be at-tached by running screws through the rail tops into their ends. You might be tempted to toenail the balus-ters on the underside of the top rail to eliminate the

unsightly screw on top, but this doesn't make an ade-quate connection.

You can improve the appearance of this railing by adding a 2x6 cap rail over a flat 2x4 top rail. The wider cap can fully cover the end grain of the posts (pro-longing their lives) and hide the screws that secure the balusters to the top rail.

**Building the balustrade**   Rather than attaching rails to the posts and then the balusters to the rails, I often build a "ladder" of the rails and balusters and then at-tach the whole unit to the posts. This job will be all that much easier if the posts are evenly spaced. I like to fit the bottom rails into ½-in. dadoes cut or routed in-to the sides of the posts. I cut the dadoes with a router and homemade jig, as shown in the drawing on the facing page. This method requires that the bottom rails be slightly longer than the top rails. Creating this joint takes some extra time, but it allows the railing to take the weight of someone who wants to stand on it for a better view. It's easiest to cut the dado in the post be-fore the post is installed. The top rail can be simply butted into the posts and attached with "toe screws."

To make the ladder, first cut a pile of balusters to length with a power miter box or a table saw. Whichever you use, set up a stop-block jig (shown in the bottom photo on the facing page) for quick and accurate repetitive cuts. You can roughly estimate the

You can create a stronger railing by inserting the bottom rail into a dado cut in the post.

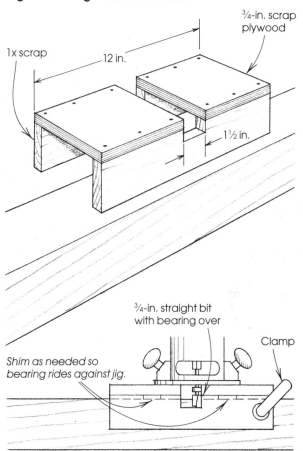

1x scrap

12 in.

¾-in. scrap plywood

1½ in.

¾-in. straight bit with bearing over

*Shim as needed so bearing rides against jig.*

Clamp

number of balusters by dividing the perimeter length of the deck by the distance between centers of the balusters. Leave out one baluster for each post and don't figure openings in your calculation.

Next, measure the distance between the posts and cut the top and bottom rails to their respective lengths. Then mark a layout for the balusters on the rails as described on p. 128. Place the top rail on the bottom rail and drill pilot holes through each simultaneously. Be sure to drill square to the face of the rails. Use a drill bit the same size as the diameter of the screw threads. This way, the threads won't catch in the rail, but they will pull the baluster tight. The screws should be driven into the center of the baluster ends. You can mark the centers by drawing diagonal lines connecting the corners of the ends.

Calculating the spacing for the balusters can be done with careful mathematics or a simple empirical method. I used to favor the precision offered by the math, but I've found that people rarely notice slight differences in baluster spacing. I'll tell you how to do it both ways.

The first step is to determine how wide the space needs to be between balusters. Codes often specify a 4-in. maximum, so if you're using 1½-in. balusters on this 4-in. grid, then each baluster/space unit takes up 5½ in. Now measure the total length of the rail between the posts and subtract 4 in., since there will be

Cut balusters and posts to length on a power miter saw with a jig with a stop block to ensure accurate cuts.

one more space than there are balusters. Divide the total by 5½, and round off the fractional result to the next highest whole number. This will equal the number of balusters needed. Now multiply the number of balusters by the thickness of each baluster (1½ in.) to arrive at the total width of all balusters. Subtract this figure from the full length of the rail to get the total width of all the spaces. Divide this total spacing by the number of balusters, and the result is the amount of space between balusters. The number will usually be a bit less than 4 in., but if your math was accurate it will never be more. Add this spacing to the thickness

of a baluster to get the center-to-center distance between balusters.

For example, if the total length between posts is 57¾ in. (or 57.75 in.), subtract 4 in. for a total baluster-space distance of 53.75 in. Divide this by 5.5 to get 9.77, which rounds off to 10 total balusters. Ten balusters will measure a total of 15 in., which, subtracted from 53.75, results in a total spacing of 38.75. Divide the total spacing by the total number of balusters to arrive at a 3.875-in. space between each baluster. Add this to 1.5 in. to get 5.375 (or 5⅜ in.) from baluster center to baluster center.

To lay out this distance, measure for the first space (3.875 or 3⅞ in.) plus half the thickness of a baluster (.75 or ¾ in.) and make a mark. The rest of the layout (5⅜ in. o. c.) can be done quickly using dividers or a calculator with a constant button to add the spacing to the sum of spacings before it.

A simpler method is to use a spacer. First find the number of balusters as described above. Cut a spacer board the combined width of a single spacing and baluster (5½ in. in this example), then find the center of the rail. If you need an even number of balusters, the center of the rail will be the center of a space; if you have an odd number of balusters, the center of the rail will be the center of a baluster. With this information you can use the spacer to lay out the centers of the balusters. If you are attaching balusters to vertically oriented rails, you can use a spacer to space and align the balusters properly. However you do it, you want to try to keep approximately the same spacing between each baluster.

**Balusters can be installed quickly and accurately by using a spacer jig. For best results, begin installing balusters at the middle and work back toward the posts.**

**Railing ends should be given a little extra attention. Edges can be rounded and corners can be clipped for a smoother transition.**

## Variations on the Wood Railing

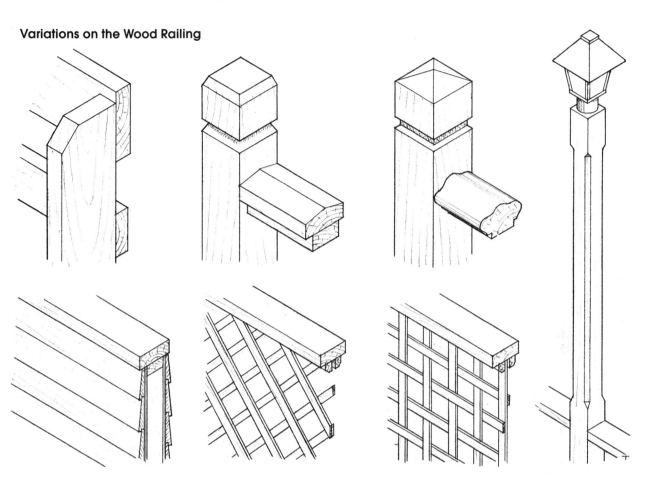

After the ladder of rails and balusters is built, slide it into place. The lower rail should fit into dadoes in the post. Fasten the rails to the posts with galvanized screws or finish nails, either through the top if it's going to be covered with a cap rail or underneath so they won't be too visible.

**Cap rails** The cap rail should be made with straight, dry, attractive boards. The cap can be a flat piece of wood, but if you're concerned about rot you may want to bevel the top surface. The easiest way to bevel a board is to run both sides through a table saw with the blade set at a slight angle. One disadvantage of a beveled cap rail is that you can't set anything on it.

Whether you choose to bevel the top of the cap or not, I like to round them off with a ⅜-in. round-over bit to add visual and tactile appeal (and to cut down on splinters). Don't try to add too much fine detail with a router as softwoods tend not to hold crisp edges over time. The cap rail should be attached to the upper rail by screws driven from underneath. This hides the screws and puts them out of the rain.

I usually bevel the butt joints in the cap rail and try to have them fall over something solid so that there's plenty to nail to. Beveled butt joints can open in time, but they don't look as bad as square butts when they do. All corner connections on the railing should be mitered and fastened with finish nails. Where the top cap ends, at a stair for example, it can be given a little extra treatment by cutting out a semicircle or nipping the corners before routing.

**Options** There are endless variations on this basic railing system. Top and bottom rails can be on edge or flat. Balusters can be nailed to the face of the rails or they can butt into them. The top (and sometimes the bottom) of the balusters can be sandwiched between horizontal 1x3s, either eliminating or in addition to the top and bottom rails. Another popular technique is to separate the cap rail from the top rail.

Railing designs can use almost any material that is strong enough to provide the necessary security. The drawings above and on p. 130 should give you some inspiration. Look around, be creative.

## Curved Railings

If your deck is curved, then your railing should be curved as well. There are two basic ways to do this on a wood railing, depending on which part of the railing is curved. A railing with a wide, vertical top rail can be constructed with laminations, much like those used on the curved rim described on pp. 100-102. The quickest way to do this is to install the posts and then laminate the rail to them. Since the rail won't be as wide as the curved rim, it can be put together in one step, flat on the ground. Just make a sandwich of wood and glue, then turn it on edge and clamp it to the bottoms of the posts, starting at the middle and working toward the ends. You'll probably want to have a helper when you do this job. The sandwich can be held together by tying strips of old rubber inner tubes around them. Be sure to spread a sheet of plastic over the deck and posts so that you don't get glue all over them.

Use a lot of clamps, one every 3 in., with wide blocks under each to spread the force and help get both edges squeezed shut. Once the glue is dry, the top and bottom edges can be cleaned up with a belt sander, rounded over with a router and bolted to the post. The glue may set quickly, so you'll want to make sure everything was thought out and cut properly before you begin. You may find that the posts are spaced too far apart for a smooth curve, for example, in which case you'll need to add extra blocks at the edge of the deck to clamp to.

Curves in horizontally oriented rails, such as cap rails, are made by first building a faceted turn with small sections of straight-sided lumber that are mitered and glued together. Then a smooth curve is cut on the inner and outer faces of this angular bend. To do this, first lay out the curve on some plywood on the deck (see the drawing at left on the facing page). Use a nail to indicate the center of the circle and draw the inner and outer profile lines of the finished curve on the plywood. Then, using railing stock, build mitered segments to fit over the drawn lines, taking care never to leave the final curve lines exposed. The segments are best joined using a biscuit joiner. After the glue is dry, lay out the curve again, this time on the segments, and

**Alternatives to the Wood Railing**

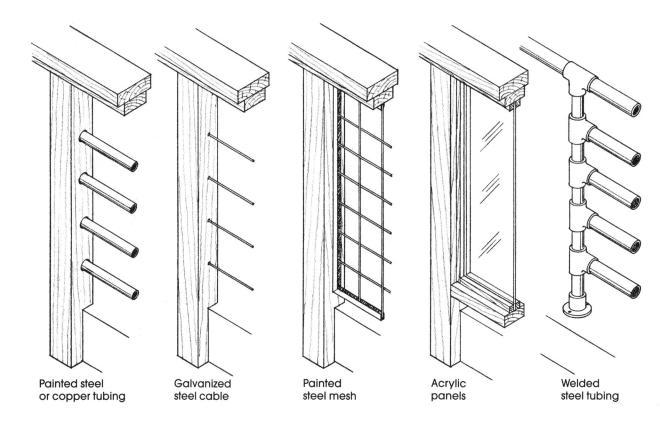

Painted steel
or copper tubing

Galvanized
steel cable

Painted
steel mesh

Acrylic
panels

Welded
steel tubing

## A Curved Railing

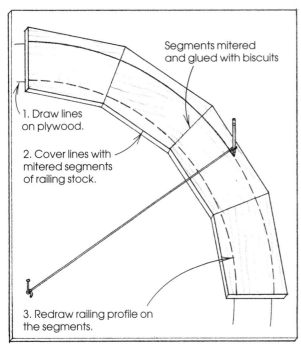

Segments mitered and glued with biscuits

1. Draw lines on plywood.

2. Cover lines with mitered segments of railing stock.

3. Redraw railing profile on the segments.

4. Cut railing stock along profile lines.

## Boxed Post Sleeve

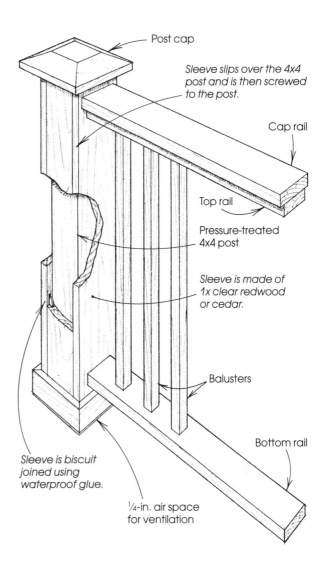

Post cap

*Sleeve slips over the 4x4 post and is then screwed to the post.*

Cap rail

Top rail

Pressure-treated 4x4 post

*Sleeve is made of 1x clear redwood or cedar.*

Balusters

Bottom rail

*Sleeve is biscuit joined using waterproof glue.*

¼-in. air space for ventilation

cut it with a jig saw. The butt joints in this type of rail are inherently weak and need to be supported, either by balusters or some other component of the railing system.

## Boxed Posts

Sometimes a simple 4x4 post just isn't large enough or fancy enough for the deck. You can use a 6x6, but these tend to develop cracks over time. A cracked post may not matter on a rustic-style deck, but it won't look right on a more refined style. The solution is to install a pressure-treated 4x4 as usual and then cover this with a square sleeve or box. The inside dimensions of the sleeve should be slightly larger than the 4x4 to allow a little room to adjust the sleeve for plumb when installing it over the 4x4.

This method works best with posts that run through the decking. It's too difficult to fashion a box with a closed bottom to cover the end of the post if the posts

are bolted to the outside of the rim. The box should be cut square on the bottom and nailed or screwed to the post. Be sure to leave a ¼-in. gap at the bottom for ventilation.

The sleeve should be made of rot-resistant wood and joined with biscuits and waterproof glue. Moldings and trim can be added to match the style of the building. A separate ornamental cap (called a finial) or a continuous cap rail are necessary to hide the construction. You can make your own caps, or they can be purchased in a wide variety of styles.

# Stairs, Benches and Other Accessories

## Chapter 7

If you want easy access to the area surrounding a deck, you'll need a set of stairs. Large decks may need more than one set. The location of the stairs will usually depend on the relative importance of the adjacent area and traffic patterns.

Exterior stairs should be built a bit larger than interior stairs. For one thing, you usually don't face the same space constraints. But there's also a difference of scale—small steps surrounded by a large house and deck look skinny. I usually aim for stairs with a width of 4 ft.—3 ft. would be the absolute minimum. Stairs can be built in a variety of shapes, with straight runs, U-shaped and L-shaped being the most common. Landings can be expanded into minidecks. You might

The landing in front of this house could have been a simple 4-ft. square dropping to an 11-in. deep stair. But by building this generously wide and deep structure, the 'stairs' become a comfortable and usable level change.

even think of expanding the stairs into a series of interesting level.

Exterior stairs need to feel grand, not hurried. For this reason I like to keep the rise of each step to about 6 in. I also aim for a tread width between 12 in. and 16 in., which creates a sense of cascading decks rather than simple stairs. Wider tread widths also make for safer footing when the stairs are covered with ice or snow. It's simplest to make the tread size compatible with a multiple of the boards you're using—three 2x6s plus gaps would yield a tread of just over 16½ in. On the other hand, stairs with 12-in. treads won't take up so much room if you need to drop 8 ft.

Wide treads and shallow risers will make the stairs longer, which is usually no problem outdoors, but it does influence construction somewhat. Stringers used to support these wide treads will be at more of a horizontal angle than usual. When there's more than a few treads, the stringers will start feeling bouncy. Long runs of stairs with these types of riser/tread-width ratios should be broken into sections with shorter stringers connecting landings or different levels. Alternatively, plan on adding support under continuous stringers or narrow the tread width to make the stringer more vertical.

Stairs are usually built with three stringers (also called carriages), which are notched in a saw-tooth manner. The bottom of the stringers rest on a pad of concrete or gravel. If the top tread of the stringer is flush with the decking, a vertically cut face on the top of the carriages butts into the rim joist. But sometimes the rim joist will serve as the top riser, and then the top tread of the stringer will be one riser down from the decking. Then the stringers will have to butt into

A typical deck stair has overhanging 2x6 treads fastened to three stringers. The risers are usually left open. Here, the stringer is nailed through the back of the rim joist.

## Covering the Stringers

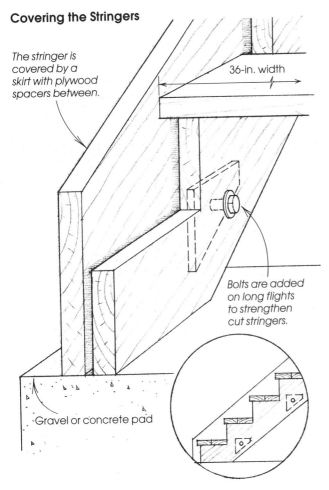

The stringer is covered by a skirt with plywood spacers between.

36-in. width

Bolts are added on long flights to strengthen cut stringers.

Gravel or concrete pad

some added framing or extend up underneath the deck and be bolted to the framing. On exterior stairs, the vertical space between treads is usually left open, but a finish riser could be added for decorative purposes.

Treads usually overhang the stringers on three sides for a traditional look, but unnotched skirts can be added to the sides of the stringers to cover their ends. These skirts are usually 2x material, which should be spaced away from the cut stringer with plywood spacers. This technique also helps strengthen cut stringers.

When planning stairs, give some consideration to the direction in which the tread boards will run. If possible, run the tread boards in a direction different from the main decking. This can help visually alert people to the approaching drop. This is particularly useful when small level changes occur unexpectedly in the middle of a section of the deck (see the photo on p. 112).

When building stairs with one or two risers to connect two levels, it might be easier to build boxes rather than to cut stringers. Built as separate units, the boxes are then stacked on top of each other, with the largest on the bottom, to reach the right height. (For more information on level changes, see pp. 96-98.)

Stairs can also be built by stacking 6x8 pressure-treated timbers on top of each other. You can also avoid cutting stringers by building stairs with treads resting on angle brackets that are bolted to side boards. I don't care much for this type of stair as the brackets and all of their bolts have a way of working loose over time. If you want to cover the ends of the treads, just build the stairs with the cut stringers and apply some nice-looking trim over the exposed stringers.

Treads are usually made from the same material used for the decking. Stringers are made from pressure-treated 2x12s. A cut stringer has a lot of exposed end grain, which should be brushed thoroughly with preservative before putting the treads on. You could also flash the stringers under the treads with Deck Seal (see the Resource Guide on p. 150).

## Codes

Although some inspectors may be lax about enforcing code requirements on exterior stairs, don't count on it. Most code requirements for stairs are easy to meet and make good sense. Common requirements are for a maximum rise per step of 8 in. and a minimum run of 9 in. Following these dimensions to the letter will result in a steep stair, which is not appropriate for a deck. For safety reasons, some codes some have changed to a 7-in. maximum rise/11-in. minimum run rule, which

## Common Code Requirements for Stairs

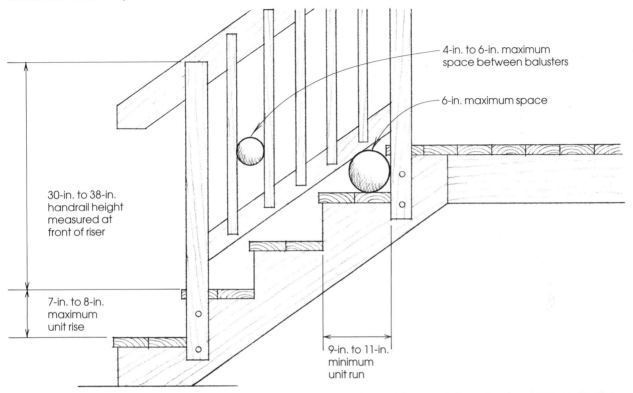

4-in. to 6-in. maximum space between balusters

6-in. maximum space

30-in. to 38-in. handrail height measured at front of riser

7-in. to 8-in. maximum unit rise

9-in. to 11-in. minimum unit run

is a common requirement in commercial construction. I recommend that all exterior stairs should be at least 36 in. wide, even if narrower ones are allowed for infrequently used areas.

Handrails are usually required when there are three or more risers on the stairs. The vertical measurement (made directly above the cut for the riser) from the top of the tread to the top of the railing must be between 34 in. and 38 in. (30 in. to 36 in some codes). Balusters must have the same spacing as on the railing, usually 4 in. However, the triangle formed by the tread, riser, and the bottom rail where the balusters are attached can usually have a 6-in. space.

## Layout

To lay out a stringer you need to determine the total rise, which is the vertical distance from the finished upper surface to the finished lower surface. Divide this distance by the rise per step that you're aiming for (I like 6 in., even if the code says I can make it 7 in.) and round off the result either up or down, depending on whether you want a rise that is slightly less or more

**When measuring the total rise on exterior stairs, it's important to measure to the area on the ground where the bottom of the stairs will rest. If you measure straight down from the deck, your figure may not account for irregularities in the ground.**

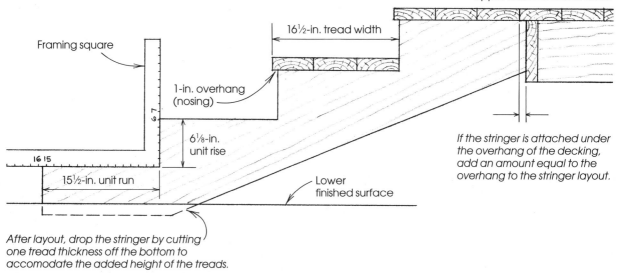

Framing square

Upper finished level of deck

16½-in. tread width

1-in. overhang (nosing)

6 7

6⅛-in. unit rise

16 15

15½-in. unit run

*If the stringer is attached under the overhang of the decking, add an amount equal to the overhang to the stringer layout.*

Lower finished surface

*After layout, drop the stringer by cutting one tread thickness off the bottom to accomodate the added height of the treads.*

than your goal. This gives you the number of risers. Then divide the total rise by this number of risers and the result is the rise per step (or "unit rise"). This measurement represents the distance from the top of one tread to the top of any adjacent tread. For example, if the total rise on a deck is 18⅜ in., and you wanted 6 in. risers, you would divide 18.375 by 6. The result, 3.06, should be rounded to three risers. Then divide 3 into 18.375, for a rise of 6⅛ in. for each step.

Next, you need to decide the run per step ("unit run"). Add to this the overhang ("nosing"), and you get the tread width. For example, if you're using two 2x6s for treads, the tread width would be about 11 in., which includes the gap between them when the boards are dry, but the run would only be 10 in., allowing the tread to overhang the stringer edge by an inch.

Now transfer the rise and run dimensions to the stringers. Use 2x12 pressure-treated lumber that's straight, uncupped and free of large knots. I use three stringers for stairs up to 48 in. wide. Set the rise and the run on a framing square using either stair buttons (as shown at left on the facing page) or by clamping on a 1x4. Then slide the square along the edge of the uncut stringer and mark the notches to be cut.

If the top tread of the stairs is going to be flush with decking (as shown above), you should lay out the total number of unit rises as calculated above. If the rim joist is going to serve as the top riser, lay out one less than the total number of unit rises on the stringers.

After the stringers are laid out, there are two important adjustments that need to be made. If you were to cut and install the stringer as laid out, you would find that the bottom step would be too short by the thickness of one tread and the top step would be too small by the same amount. You can easily solve both problems by simply removing the thickness of one tread from the bottom of the stringer. This is called "dropping the stringer."

If the stringers will be hung on the rim joist (mine usually are), you'll want to make the top horizontal cut wider by the amount that the deck boards overhang the rim joist. Then the stringers can fit under them and you'll still have room for the top treads to be as wide as, and look the same as, all the others. You can cut the stringers with a circular saw, but don't overcut the layout lines just to finish the cut. This weakens the stringer and looks tacky as well. Instead, use a handsaw to finish the cut.

## Installing the Stringer

The bottoms of stringers need to rest on something solid and well drained. A simple technique is to excavate a 6-in. deep footing and fill it with well-compacted gravel. Then, before nailing the stringers to the deck, turn them over and nail two pressure-treated boards across the bottom. Once the assembly is installed these boards will help to distribute the load around the gravel.

Sometimes gravel looks too casual, in which case a concrete pad might be better. Technically, a concrete pad should have a foundation below the frost line, but if the pad rests on a 6-in. layer of gravel in well-drained soil it shouldn't experience frost heave. It's difficult to know where to locate the pad until the deck has been built, so this requires doing some concrete work for a second time. The pad can be big enough to provide a landing at the bottom and a base for the stairs.

I know of no anchors made specifically for attaching the bottom of the stringers to concrete, so you'll need to improvise. A piece of galvanized angle iron will work when bolted to the concrete and the stringers. The stringers can also be secured with a pressure-treated kicker notched into the bottom of the stringers. The kicker is nailed to the stringers and bolted to the concrete. By careful measuring it will be possible to locate and set the bolts needed for the anchors while the concrete is still wet.

Stair buttons are set on the framing square to the unit rise (left) and unit run (right). Then the stringer can be laid out and cut.

You can save a little time by using the stringer you cut first as a template to lay out the others.

## Anchoring the Bottom of the Stairs

### On Gravel

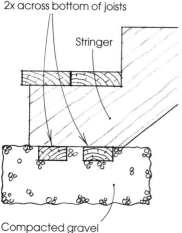

Pressure-treated
2x across bottom of joists

Stringer

Compacted gravel

### On a Concrete Pad

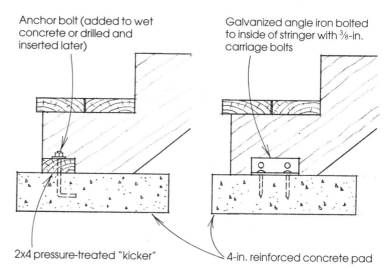

Anchor bolt (added to wet concrete or drilled and inserted later)

Galvanized angle iron bolted to inside of stringer with ⅜-in. carriage bolts

2x4 pressure-treated "kicker"

4-in. reinforced concrete pad

## Anchoring the Top of the Stairs

### Attaching the Stringers When the Rim Joist Serves as the Top Riser

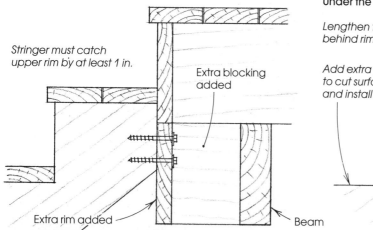

*Stringer must catch upper rim by at least 1 in.*

Extra blocking added

Extra rim added

Beam

### Attaching the Stringers Under the Deck

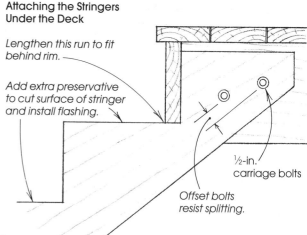

*Lengthen this run to fit behind rim.*

*Add extra preservative to cut surface of stringer and install flashing.*

½-in. carriage bolts

*Offset bolts resist splitting.*

**These stringers are anchored solidly to the rim joist using regular joist hangers that have been clipped. They could also be fastened with lag bolts or, on short flights, nails driven through the backside.**

Anchoring the top is more critical. I try to design my deck stairs so that the tops of the stringers rest fully against the rim joist. This makes it easy to attach the stringers. On short flights with just three or four steps, you can nail through the backside of the rim into the stringers with several 20d nails per stringer. On longer flights or with a rim thicker than 1½ in., several lag bolts will be better. It's a good idea to reinforce this connection by using some galvanized framing anchors or joist hangers clipped in half. Nail these to the stair side of the rim but only under the tread area where they're not highly visible.

When the rim joist functions as the top riser, it can't serve as a support for the ends of the carriages. Instead, an extra piece of joist stock should be added under the rim to hold the stringers and should be braced up to the main joists or against a beam. The stringers are then attached to the rim. The stringers can be left extra long on their top end to extend up under the deck; bolt them either to existing joists or to added blocking.

## Stair Rails

A handrail for an exterior stair is a practical necessity even if it's not always required by code. It's usually built in the same style as the railing for the deck. Posts can be notched and bolted to the stringers much the same as they were to the rim joist. In order to get the maximum surface to bolt the post to, I place the posts even with the front riser cut. If the treads overhang the stringers on the sides, the treads will need to be notched to fit around the posts. Make the notch in the post high enough so that fits over the treads and the stringer. You can use ⅜-in. or ⁷⁄₁₆-in. carriage bolts to fasten the posts to the stringer. Try to space the posts on the stringer the same distance apart as on the rest of the deck or, if necessary, less to get them evenly spaced along the stringer.

One of the most common railings seen on exterior stairways uses a flat 2x6 as a cap rail. While this is fine for a deck railing, it may well be a code violation when used as a stair railing. Handrails on a stair serve an important safety function, and to serve that function

This simple T-shaped handrail was made with a 2-in. wide board nailed on top of a 2x6.

they must be easy to grasp. A 1½-in. to 2-in. round handrail is the easiest to grasp, but variations are allowed. For example, I made a T-shaped handrail for the Morrises' deck by attaching a 2-in. wide board on top of a 2x6 that was installed on edge (shown above).

The junction of stair rails and deck rails always presents an awkward dilemma. There are three normal solutions. First, the stair rail can butt into a post holding up the deck rail, but at a lower level. Second, the post can be located so that the stair rail and deck rail intersect on one post to form a continuous railing. (Finding the exact location for this post will require careful planning and is best done by first building a prototype in place with scrap lumber.) Finally, the deck rail and stair rail can both start and stop on their own independent posts (see p. 128).

## Ramps

I built a ramp on the Morrises' deck because they had a friend who used a wheelchair and they wanted to make his visits as barrier-free as possible. Once the ramp was in place, however, they found it facilitated moving bicycles, garden carts, appliances and many other items around and into the house. Ramps are built like small sections of a deck, with joists and decking of the same type used for the rest of the deck. Span tables for joists will give the size required for the length of ramp.

A ramp can also be built as a boxed unit and then bolted to the main deck. This would make it easy to remove the ramp at a later date without much visible effect on

The joists on this ramp are attached to the rim with clipped joist hangers. They will be covered with 2x6 decking. Ideally, the bottom of the ramp would be supported by a concrete pad.

the deck. Joist hangers would be wise reinforcement where the joists butt the end of the box. Ramps should have a maximum slope of 1 in 12 — 1 in 15 would be better. They should also be at least 4 ft. wide to accommodate wheelchairs. Finally, they should have level landings at the top and bottom.

The joists of a nonremovable ramp can be anchored to the rim joist of a deck in the same way as stair stringers, with nails or bolts and joist hangers. If the ramp rises more than 6 in. or is longer than 6 ft., it should be equipped with a railing, which should be built to the same standards as the stair railing.

The bottom of the ramp is supported like a stairway. It can rest on a gravel pad with several pressure-treated boards across the bottom of the ramp's joists to provide extra bearing surface. However, loose gravel may not be appropriate if the ramp is used for a wheelchair. A better solution would be a concrete pad, which could be sloped to provide a smooth transition.

## Seating

Built-in benches are a good way to provide seating for a lot of people at social gatherings. Permanent benches near doors can be a place for changing muddy shoes or resting the groceries while you unlock the door. Seats built into the corners of a deck can foster private conversation. Attractive built-in seating can add character to a deck by filling in a monotonous section of railing. On a low deck that doesn't require a railing, a perimeter bench is a useful and appealing addition.

That said, I should add that I'm not a big fan of built-in seating on decks. For one thing, it's often less-than-comfortable seating. And, for some reason, benches are frequently built along the rail facing the house, which places the users' backs to the view. Bulky benches can also disrupt the view from a deck. I usually advise people to limit built-in seating to the sides or corners of a deck, unless they're going to need seating for a large group of people on a regular basis, in which case benches can be cheaper than buying a lot of furniture.

Good outdoor furniture is a much more comfortable and versatile approach to sitting on a deck. It can be moved as needed to get into or out of the sun, depending on your desire. It can be gathered around a table for a meal and then moved against the house to enjoy the sunset. It can be dragged out into the yard and, if it blocks your view, you can simply remove it. Plans for building outdoor furniture—Adirondack chairs or picnic tables, for example—are readily available in woodworking magazines and at lumberyards.

If built-in benches do fit your design, there are several ways to build them. A section of the railing can be modified to form an angled backrest for the bench, but this method creates a mismatch with the rest of the deck's rail. You can also build a bench without a separate back, or build a separate backrest and use the main

**The supports for this bench are built using the jig shown on the facing page. The supports are bolted to posts and the bottoms are toenailed to the deck. The seats and back rests are made with decking material.**

railing and posts just for support. This method requires a bit more work and material than the first, because you must plan ahead to get the posts where you need them for the bench supports.

Built-in benches need legs spaced about every 4 ft. for adequate support. One of my favorite techniques requires an angled support system for the bench, for which I make a simple jig to mass-produce the supports. The jig uses a few scraps of lumber nailed to a piece of plywood. It allows quick and consistent alignment of the parts in each support. The bench support involves only three pieces—a long back, a horizontal base for the seat and a short front leg. The pieces are precut, placed in the jig and bolted together. Then they're attached to an existing deck post. Finally, they're covered with decking material to form a seat and back.

The seat should be slightly higher in the front than the back. The width of the seat planking should be a multiple of the material being used, say three 2x6s for a total of 16½ in. These can be spaced a little away from the back support to provide extra width if desired. The horizontal support for the seat should be slightly shorter to allow the seat to overhang. Two or three 2x6s also work well for the backrest. Since the seat is flat, leave a large space between the backrest and the seat to make the bench more comfortable.

## Overhead Protection

If you're building a deck, chances are that you want to be out under the sun and stars. But sometimes it's nice to be able to find some shade. A trellis is a common way to do that. Trellises can be built with single or multiple layers of wood to achieve the desired degree of shading. When covered with seasonal greenery, a trellis can provide a lot of shade in the summer and allow the winter sun to shine through (see the drawings on p. 12). Canvas awnings can also be installed on a deck, which can be rolled out as needed. A large cafe-style umbrella or two is another idea for putting shade exactly where you want it.

As mentioned in Chapter 1, you might consider building a permanent roof over a section of the deck. The roof can be made with translucent fiberglass panels, or the main house roof can be extended over the part of the deck. If desired, the sides could be closed off with removable screen, plexiglass or lattice panels.

## Planters

Planters are a natural deck accessory—they can be as important as pictures on a blank interior wall. They can be built in a variety of shapes and sizes to fit their location, and they can be built on casters that allow their location to be changed. Pairs can be placed at both sides of a stairway or door to define an access or delineate changes in levels. Planters can serve as dividers or barriers, actually replacing sections of a railing.

Planters are simple wooden boxes. They can be built with a false bottom so only a minimum of soil is required at the top for some flowers, or they can be sized to hold a plastic liner. They should be designed to allow maximum air circulation, especially between the bottom of the planter and the deck. The bottom

### Jig for Bench Supports

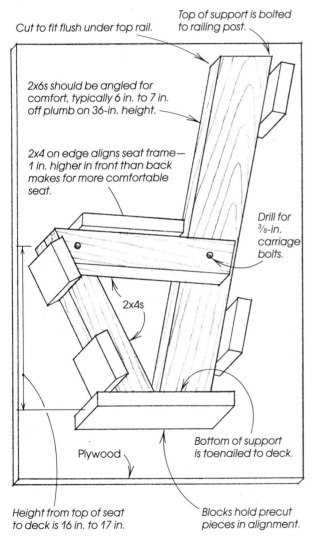

Cut to fit flush under top rail.

Top of support is bolted to railing post.

2x6s should be angled for comfort, typically 6 in. to 7 in. off plumb on 36-in. height.

2x4 on edge aligns seat frame—1 in. higher in front than back makes for more comfortable seat.

Drill for ⅜-in. carriage bolts.

2x4s

Plywood

Bottom of support is toenailed to deck.

Height from top of seat to deck is 16 in. to 17 in.

Blocks hold precut pieces in alignment.

A few of these planters can make a grand statement on a deck. This planter is constructed with the same wood used for the decking. It has a false bottom and allows for good ventilation.

## Building a Planter

*Note: Adjust overall dimensions to suit*

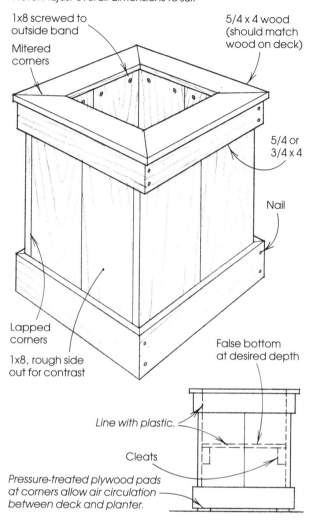

1x8 screwed to outside band

Mitered corners

5/4 x 4 wood (should match wood on deck)

5/4 or 3/4 x 4

Nail

Lapped corners

1x8, rough side out for contrast

False bottom at desired depth

*Line with plastic.*

Cleats

*Pressure-treated plywood pads at corners allow air circulation between deck and planter.*

should be sturdy enough to handle the load, and it should allow for water drainage. One of my favorite planter designs is shown in the photo above and in the drawing at right.

## Lighting

I rarely worry about incorporating lighting into the decks that I build. But that has a lot to do with the fact that I build in Alaska, where the sun shines almost round-the-clock in summer, and in winter...well, let's just say we don't use our decks much in the winter. If your deck is going to get a lot of use in the dark, you'll certainly want to consider getting some light onto it. Floodlights and spotlights are good for maximum illumination. A pair of 150-watt to 250-watt reflector floodlights mounted as high as possible should be enough for most situations. Spotlights and floodlights can be used most creatively when they "wash" an object rather than garishly illuminate it. Fixtures that shine from the ground up or through the mottling branches of trees can provide dramatic effects. Go out-

side with some portable lights and try out different ideas before you actually make anything permanent.

Your local code will determine whether or not you can do your own electrical wiring. Electrical requirements can be as numerous and confusing to the novice as they are important, so don't hesitate to get professional help. Exterior wiring requires the use of watertight fittings and boxes as well as conduit to prevent any wires from being exposed. All 120-volt outdoor circuits and objects such as hot tubs are required by code to be protected with a device called a ground fault interrupter (GFI). GFIs are important safety devices that will shut off the flow of electricity at the slightest hint of current leak to a grounded source, such as a person holding a malfunctioning tool while standing on damp ground.

What decks really benefit from, in my opinion, is more subtle mood lighting. This means lower wattage fixtures, strategically placed, that enhance the beauty of the deck and offer just enough light to find your way around. This kind of lighting works well around stairs and level changes, near doorways and along pathways. Incorporating lamp posts into railings can provide dramatic pools of light.

Low-voltage lighting works well for lighting small areas of a deck. A transformer converts the standard 120 volts from the house circuit to 12 volts, which is the same as your car. You can even use automobile bulbs and fixtures in a 12-volt deck system. Even if you aren't allowed by local code to do your own 120-volt electrical wiring, most codes will permit you to do your own 12-volt wiring. In fact, building-supply centers often carry complete packages that include low-voltage light fixtures, wire and a transformer.

Since voltage drop is significant on long runs of wire, the total length of wire on one transformer is limited to about 100 ft. Also, the transformer may be restricted to about 50 watts, which means about five or six low-voltage bulbs. You may find it necessary to use several transformers on a highly lit deck.

You might be able to handle all of your lighting needs on a deck without doing any wiring at all. Solar lights made by Siemens and other manufacturers are available with built-in solar cells. They are reasonably priced and can provide many hours of "off the grid" lighting (see the Resource Guide on p. 150).

## Under the Deck

After the building is done, the area under a deck is likely to look like a construction site. Dirt will be in piles from digging holes for the footings, and the grass will be worn and perhaps yellowed. It's unlikely that the grass will ever recover entirely since sunlight is no longer reaching it full strength.

The simplest and most frequent recommendation is to dig up the old grass, cover the dirt with black plastic to inhibit further growth, and cover this with gravel. Put a lot of holes in the plastic so that water won't collect. The gravel can be held in place with some perimeter timbers. On a low deck you may want to screen off the underside using some lattice from the lumberyard. Adding horizontal rails to the under-deck post system will give you something to nail the lattice to. Instead of using lattice, wide cedar boards can be added as trim or used to build panels that look solid.

# Finishing and Maintenance

## Chapter 8

Protecting the wood on a deck is tough work, and there are no shortcuts or cure-alls. Most finishes need repeated applications, sometimes more than once a year. You'll have taken an important first step in providing for the long-term health of a deck by designing and building it well and using quality materials. Decisions on finishes and maintenance routines should follow the same standards.

Decks face a variety of abuses—the biggest problems are water and sunlight. The sun's ultraviolet rays, which are becoming ever more intense with the weakening ozone layer, break down the cells of the wood and destroy its natural preservatives. They also attack the synthetic preservatives, water repellents and pigments contained in deck finishes, which is one reason why protection must be reapplied frequently.

But destruction by sunlight pales in comparison to what happens when the wood absorbs water and then dries out on a regular basis. The resulting expansion and contraction of the wood pulls against fasteners and causes warps, splits and cupping. This in turn lets moisture penetrate further into the wood. Freezing conditions only add to the degradation. Moisture in the wood is also one of the ingredients necessary for the growth of fungi and bacteria that are responsible for rot. Finally, decks are subject to heavy foot traffic. The purpose of a good finish is to protect the deck as much as possible from all these enemies.

In dry and mild climates, it's possible to build a deck out of good redwood, cedar or pressure-treated lumber and then leave it alone and still be assured of many years of life. Untreated redwood and cedar will turn grey over time, which some people seem to like. With their natural preservative qualities, these woods con-

tinue to survive in this condition, but can still benefit from an occasional application of finish.

CCA-treated softwood will also weather to a grey color that won't be quite as attractive as redwood and cedar, and most people find it a big improvement over the original greenish tint. But treated lumber has its natural oils tortured during the treatment process and becomes significantly less weather resistant over time unless the deck is regularly coated with a good finish.

The weathered look isn't everyone's style, however. If you prefer the "natural" look of warm brown tones, you will have to be all the more vigilant and make maintenance a priority.

If you've let a deck weather naturally for a while and decide that you would really like to make it look new again, there are a variety of bleaching products available that will remove the grey color, mildew, fungi and dirt. They can be brushed or sprayed on, allowed to sit for a while, and then rinsed off. After the deck has dried it will be close to its original color and can then be coated with the finish of your choice.

## Deck Finish Ingredients

I've tried many different brands of deck finishes. Some have performed better than others, but they all fall short of perfection. Finishes are formulated to combat one or more of the principal enemies of decks, and you may not need an all-purpose finish in your region. Learn to identify the various finishes by their ingredients—there are lots of manufacturers, but they all make a narrow range of products.

The main ingredient in all the finishes is the solvent, which is the portion that evaporates after application, leaving the solid ingredients behind. Traditional solvents in wood finishes have contained a lot of toxic substances known as volatile organic compounds (VOCs). I have mixed feelings about recommending deck finishes with high VOC levels, but until recently they were usually all that was available. Now, many states and localities have banned these old formulas, forcing the manufacturers to come up with less toxic ingredients. Even if VOC regulations haven't hit your area yet, you might be able to find "VOC-compliant" versions of some deck finishes. They generally have a higher solids content and are slower to dry, but they should give the same level of protection as the original versions. Most of the ingredients in deck finishes are toxic and should be used with care.

I strive to build a well-drained deck of good materials that requires only a water repellent. But my clients often want to retain the "natural" wood color, which requires that I recommend a reliable finish that can be added with a minimal expense.

**Water repellents**  Any finish used on a deck should contain a water repellent, which usually is paraffin. Finishes that are only water repellents are frequently sold as sealers. They will not stop the greying process and they provide a minimum of protection. As with all wood finishes, they should be applied only to dry wood. As a general rule, you'll know that your deck is repelling water if the water beads after a rain. If it does not bead, then the wood is absorbing the water, which means that it's time for another application. Water repellents are good for all types of wood, including pressure treated.

If a water repellent is all you need, you might want to make your own by mixing equal amounts of boiled linseed oil and solvent (preferably turpentine, although mineral spirits or paint thinner would do). This finish will darken the wood. To enhance the mixture, the Forest Products Laboratory suggests first adding some flaked paraffin wax (1 oz. per gallon) to the carefully warmed solvent before adding the oil. I would only do this in a double boiler on an electric stove. Apply it like any clear deck finish, and if you leave any sticky puddles behind, they can be thinned out by scrubbing with solvent. Like most water repellents, it will need frequent reapplication.

**Preservatives**  Common preservatives in deck finishes are fungicides, mildewcides and insecticides—most all-purpose deck finishes contain these ingredients. You can help minimize the need for such preservatives by using good wood and keeping it dry. You can buy small cans of preservative to brush onto the cut ends of pressure-treated wood (this is a requirement to keep any warranty in effect).

**UV blockers**  To prevent the wood from absorbing the full blast of the sun's rays, many finishes contain suspended particles of pigment or other ingredients that reject or disrupt UV degradation. Keeping out the UV rays will help maintain the wood's natural color. Some clear finishes, which have no pigment, contain organic chemicals that function similarly to sunscreen for skin. These chemicals absorb the UV rays and convert them to heat before they can destroy any wood. Other clear finishes contain finely dispersed metal particles that reflect the sun's rays. These ingredients will wear out or rub off in a year or two, which is why they must be regularly reapplied. Pigmented products, such as tinted clear finishes or the more heavily colored semi-transparent stains, rely on the particles of pigment to do the UV absorbing and last longer than chemical blockers.

**Resins and oils**  Linseed oil and alkyd resins give finishes their amber color. They harden and seal the wood against wear and moisture. The same types of resins are found in varnish as in deck finishes, but in different concentrations. The resins keep the wood looking wet and colorful after the solvent has evaporated.

## Clear Finishes

The term "clear finishes" describes a variety of products, from colorless water repellents to the amber or lightly tinted penetrating oils. The most basic clear finish is a simple water repellent. Thompson's Water Seal is a well-known version. It keeps water out of the wood while allowing it to turn grey. Preservatives added to these basic formulas can protect against rot and mildew. These types of finishes are particularly useful on non-pressure-treated wood. Cuprinol's Clear Wood Preservative is an example.

The finish that I use the most is a clear, all-purpose type. These amber liquids contain water repellents, preservatives, UV blockers and a good amount of oils and resins. They help retain the wood's natural color, and they can be used to enhance and darken an already greyed surface. I have used both Flood's Clear Wood Finish and Penofin with success, but I've found that they need to be reapplied every two years. To increase the effectiveness of this type of finish, manufacturers have added small amounts of pigment. These lightly tinted versions are the latest generation of deck finishes, and I recommend them over other types. The tint comes in a variety of shades that imitate redwood, cedar, lightly weathered wood and even pressure-treated green. The pigment adds extra protection and evens out variations in the color of the wood. These finishes will darken the wood for a week or two after application. Many manufacturers make their products in both the tinted and untinted versions. I've used Behr Natural Seal Plus (also know as Natural Wood Finish), and a similar product is Amteco's Total Wood Preservative or their newer, VOC-compliant version, Total Wood Protectant (for a listing of finish manufacturers, see the Resource Guide on p. 150).

Tinted oil finishes go well on redwood and cedar. They can also be used to change the color of pressure-treated wood if it is first allowed to weather for a year with only a water repellent.

## Semi-Transparent Stains

Semi-transparent stains are moderately pigmented oil-base finishes that provide color to wood while still letting the grain show through. The higher pigmentation provides for better UV protection than clear finishes, but I choose a semi-transparent stain only when I want to change the color of the wood. And because I usually build cedar decks, I don't often want to hide the beauty of the cedar. These stains are better for masking the color of pressure-treated wood decks. The color in semi-transparent finishes may tend to show wear in heavily traveled paths, but they'll probably protect against UV degradation twice as long as a clear finish. Some semi-transparent stains are not recommended for use on decks, so be sure to read the label.

## Solid-Color Finishes

Solid-color finishes include paint and opaque exterior stains. I don't recommend them for decks. Because they build a colored layer on top of the wood surface, they quickly start deteriorating from traffic and the elements. Further, to refinish a painted deck you have to remove all of the loose paint first, which can be a major job. This isn't to say that a solid color can't make a deck look great, but if you feel you must go this route, I'd suggest using it only on covered areas or vertical surfaces, such as railings.

## Other Finishes

If you've ever admired the beauty of a spar-varnished wooden boat and thought about duplicating the effect on a deck—forget it. It won't last long, and it requires a lot of work to renew. A varnished deck would also be quite slippery.

Exterior penetrating oils are available from Watco and Varathane. These are thin-bodied, easy-to-apply finishes that require frequent applications. I find them more suitable for deck furniture and window and door trim than for decks.

There are several sources of low-toxic and non-toxic exterior wood finishes that you may want to investigate (see the Resource Guide on p. 150). These finishes contain linseed oil, wood pitch, citrus thinners and other plant-derived substances. I have been pleased with Livos interior paints, but I haven't used the exterior finishes. My suspicion is that they're going to require frequent application. They're also quite expensive, perhaps three to five times more costly than the more mainstream products, and they often can be bought only by mail order.

## Applying the Finish

The wood should be thoroughly dry before you apply the finish. Dry wood will soak up more finish, which will provide more protection. The surface should also be clean and free of surface oils, and the pores of the wood should be opened. Opening the pores can be accomplished by simply letting the deck age for a few weeks before finishing. You could also give the deck a quick sanding with 120-grit sandpaper on a pole sander. A new deck can be cleaned with a bleaching solution to break down the mill glaze on the new wood and allow greater penetration of the finish. Use about 1 cup of household bleach per gallon of water and rinse well afterward. The final step should be a thorough sweeping.

Before applying any finish, read and prepare to follow the application procedures recommended on the can. Most of the clear all-purpose finishes require a wet-on-wet application; that is, a second coat is applied before the first coat has dried and sealed the surface against further penetration. The objective is to saturate the wood until it won't absorb any more finish. Then after an hour or so, any wet spots are brushed in or wiped up (shiny areas will leave a tacky mess). Regardless of the finish you use, you should apply multiple coats to any exposed end grain. Try not to do all this on a hot, sunny day. The finish will dry so quickly

that you'll have trouble keeping a wet edge and getting the second coat on before the first dries. Expect coverage on the first coat to be about 150 sq. ft. per gallon, and about 200 sq. ft. per gallon on the second.

Semi-transparent deck stains usually require only one coat to avoid the build up of too much color. They should be applied liberally and with care taken to avoid lap marks.

For top-notch performance, I like to apply finish to a deck once a year for the first three years. After that, I refinish every two years. But this schedule depends on local weather and the kind of use the deck gets. Try to refinish at the first signs of greying or weathering. If you wait too long, you can renew the color of the wood following the bleaching routine described above.

**Brushes**   Brushing on the finish with a good-quality brush provides for excellent coverage, but it takes too long for me. However, I do use a brush to even out the application and mop up runs and drips after spraying clear finishes on railings and other complex shapes.

It's a good idea to brush on semi-transparent stains, however, as a brush can work the stain into the wood and make the finished appearance more even. A brush also does a good job of picking up any runs, which would be more visible with a stain.

**Rollers**   For clear finishes and water repellents on horizontal surfaces, a roller works great. It spreads the finish on fast, works it into the wood and doesn't leave puddles. If you use a roller with a long handle, you won't have to bend over to do the work. While I've found a ½-in. nap roller to work fine on new decks, some manufacturers recommend using a 1-in. nap to help get the finish into the cracks. Buy good-quality rollers—I've found that the solvent in the finish can dissolve the glues holding cheap rollers together. I don't bother cleaning the rollers—a big job would require a greater expense in solvent than a new roller would cost. You can store a used roller overnight in a sealed plastic bag.

**Pads**   A pad attached to a long handle can really work the finish into the deck. Pads can also be used to even out a sprayed application.

**Sprayers**   There are two types of sprayers that could be used to apply finish. For most deck jobs I recommend a simple and inexpensive pressure-tank type used for applying chemicals. These put out enough to do any surface quickly and are indispensable for complex areas like railings. The tank may hold a gallon or more of finish and so won't need frequent refilling, and the spray patterns are adjustable. Sprayers are available in a variety of sizes and for different uses. Although similar looking, most garden sprayers aren't intended for wood finishes. If you try to use one, you may find that the solvents quickly eat through the rubber gaskets.

Another type of sprayer is called an airless sprayer. Airless sprayers take some time to learn to use properly. In the hands of a beginner, they can result in an uneven application. They work quickly and put out large volumes of material, but they waste a lot through overspray. They're great for siding but are really unnecessary for the average deck job. The small home-owner varieties cause less of an overspray problem, but they also only hold a small amount of finish, and this slows the job down. For large flat surfaces, it's hard to beat a low-tech roller for cost effectiveness.

If you're ambitious and plan on doing the underside of your deck, then a sprayer is the only way to go. Working overhead with thin finishes is a mess, and anything that gets it on fast is my idea of good. The extra finish wasted by using a sprayer will be a small price to pay for the convenience. Spraying will also speed the work of those compulsive types who feel the need to get the finish into all the cracks between the deck boards. If you spray, you'll have to pay attention to spreading out any puddles with a brush, roller or pad, and you should look for any dry (missed) spots.

## Safety

Applying traditional deck finishes can be hazardous work. Take the time to apply the finish only where you want it—if you get some on your flowers they'll never look the same again. And keep it off yourself. Wear gloves that will stand up to the solvents if you use a sprayer. You might want to wear rubber boots as well. I usually wear a charcoal-cartridge respirator when I finish a deck unless the area is small and there's a good breeze blowing the vapors away. Don't fool yourself by wearing a cheap dust mask—it won't block any solvent fumes.

If you're applying finish to the underside of the deck you've got real problems anyway, but all I can say is to cover yourself—you'll have to work hard to keep it off your eyes, lips and skin. Wear a disposable suit and hood from a paint or safety-supply source and glasses that will protect your eyes from drips and splashes.

Rags and rollers soaked with oily deck finish are particularly susceptible to spontaneous combustion. Any used materials should be sealed in a metal container with a tight lid and disposed of according to the directions on the product label.

# Resource Guide

## Fasteners and Fastening Tools

Ben Manufacturing
21239 Cypress Way
Lynnwood, WA 98036
(206) 776-5340
*Dec-klip fasteners*

The Blind Nail Co.
Box 782
Lloydminster, Saskatchewan,
Canada, S9V 1C1
(403) 875-3325
Deck Claw fasteners
ITW Buildex
1349 West Bryn Mawr Avenue
Itasca, IL 60143
(800) 323-0720
(708) 595-3549
*Dec-U-Drive deck-fastening
system and Dec-King screws*

John Wagner Assoc.
P.O. Box 4060
Concord, CA 94524
(510) 680-0777
*GrabberGard exterior screws*

Manasquan Premium Fasteners
P.O. Box 669
Allenwood, NJ 08720-0669
(800) 542-1979
*Stainless-steel fasteners*

Maze Nails Division
W.H. Maze Co.
Box 449
Peru, IL 61354
(815) 223-8290
*Corrosion-resistant nails and screws*

McFeely's
P.O. Box 3
712 12th St.
Lynchburg, VA 24505-0003
(800) 443-7937
*Quik Drive automatic screw-driver
system*

Simpson Strong-Tie Co., Inc.
Box 1568
San Leandro, CA 94577
(800) 227-1562
*Manufacturer of deck-board ties
and wide variety of joist, beam
and post connectors*

Swan Secure Products, Inc.
1701 Parkman Avenue
Baltimore, MD 21230
(410) 646-2800
*Stainless-steel nails*

## Wood Suppliers

Champion Ridge Lumber Co.
P.O. Box 272
Whitehaven, CA 95489
*Salvaged redwood*

Greenheart Durawoods, Inc.
P.O. Box 279
Bayville, NJ 08721
(908) 269-6400
*Pau lope*

Michael Evenson
P.O. Box 202
Redway, CA 95560
(707) 923-2979
*Salvaged redwood and wood from
ecologically harvested forestland*

# Finishes

**Nontoxic**

AFM Enterprises
1140 Stacy Court
Riverside, CA 92507
(714) 781-6860

Auro-Sinan Co.
P.O. Box 857
Davis, CA 95617-0857
(916) 753-3104
*Non-petroleum-based products*

Eco Design Co.
1365 Rufina Circle
Santa Fe, NM 87501
(505) 438-3448
*Livos products*

Preserva Products
Box 744
Tahoe City, CA 96145
(916) 583-0177

**Other finishes**

Amteco, Inc.
815 Cass Avenue
St. Louis, MO 63106
(800) 969-4811

Behr Process Corp.
3400 Segerstom Ave.
Santa Ana, CA 92704
(714) 545-7101

Darworth Co.
101 Prospect Ave.
Cleveland, OH 44115
(800) 424-5837
*Cuprinol*

Flood
P.O. Box 399
Hudson, OH 44236-0399
(800) 321-3444

Thompson & Formby
825 Crossover Lane
Suite 240
Memphis, TN 38117
(901) 685-7555

# Miscellaneous Products

Heckman Industries
405 Spruce Street
Mill Valley, CA 94941
(800) 841-0066
*Deck Seal flashing*

Real Goods Trading Co.
966 Mazzoni Street
Ukiah, CA 95482-3471
(800) 762-7325
*Solar lighting*

Riecreation Products
7 Old Orchard Lane
Fairport, NY 14450
(716) 377-0070
*Deck Mate spacing tool*

T.C. Manufacturing
P.O. Box 122
Fredericktown, OH 43019
(614) 694-2255
*Lumber Jack tool*

# Trade Associations

California Redwood Assoc.
405 Enfrente Drive, Suite 200
Novato, CA 94949
(415) 382-0662

Western Wood Products Assoc.
Yeon Building
522 SW Fifth Avenue
Portland, OR 97204-2122
(503) 224-3930

Southern Forest Products Assoc.
P.O. Box 52468
New Orleans, LA 70152
(504) 443-4464

# Environmental Associations

Institute of Sustainable Forestry
P.O. Box 1580
Redway, CA 95560
(707) 923-4719
*Information on recycled lumber*

Public Forestry Foundation
P.O. Box 371
Eugene, OR 97440-0371
(503) 687-1993

Rainforest Action Network
301 Broadway
San Francisco, CA 94133
*Publishers of* The Wood Users Guide

Woodworkers Alliance for
Rainforest Protection (WARP)
P.O. Box 133
Coos Bay, OR 97420-0013
(503) 269-6907

# Index

| | |
|---|---|
| Editor | Jeff Beneke |
| Designer/layout artist | Henry Roth |
| Copy/production editor | Pam Purrone |
| Illustrator | Vince Babak |
| | |
| Typeface | ITC Stone Serif |
| Printer and binder | Arcata Graphics/Hawkins, New Canton, Tennessee |